A Hawaiian Florilegium

The National Tropical Botanical Garden, a privately supported research and education institution, was chartered by the United States Congress in 1964. Its mission is to be an international leader and resource for the propagation, study, and conservation of tropical plants and to share its discoveries broadly in order to enlist the public's help in protecting and enriching the environment. Today this work is carried out at five gardens and three preserves:

- Lawai and Allerton Gardens, Lawai, Kauai
- Limahuli Garden and Preserve, Haena, Kauai
- Kahanu Garden, Hana, Maui
- Kaupulehu and Aweni Preserves, Hawaii
- The Kampong, Coconut Grove, Florida

A Hawaiian Florilegium

BOTANICAL PORTRAITS FROM PARADISE

Mary Grierson, Botanical Artist

Text by Peter S. Green

NATIONAL TROPICAL BOTANICAL GARDEN
LAWAI, KAUAI, HAWAII

Dedicated to

Elizabeth Loy McCandless Marks
1905–1995

A leader in the initiative to obtain a Federal charter that
created the National Tropical Botanical Garden in 1964
and founder of The Loy McCandless Marks Botanical Library

Library of Congress Cataloging-in-Publication Data

Grierson, Mary 1912–
A Hawaiian florilegium : botanical portraits from paradise / Mary Grierson, botanical artist ; text by Peter S. Green.
p. cm.
Includes bibliographical references (p. 99) and index.
ISBN 0-915809-20-6 (alk. paper)
1. Botany—Hawaii—Identification. 2. Botany—Hawaii—Pictorial works. 3. Botanical illustration.
I. Green, Peter S. (Peter Shaw) II. Title.
QK473.H4G75 1996
581.9969—dc20 96-22447
CIP

Administration: Janet L. Leopold
Bibliography: Eileen Herring
Design: Paula Newcomb
Development and project management: John C. Thomas
Scientific review: David H. Lorence, Ph.D.

Distributed by
University of Hawai'i Press
Honolulu, Hawai'i 96822

Printed in Singapore

FOREWORD

A Hawaiian Florilegium is a family album of paradise, portraits of ancient voyagers who against all odds made it to the most remote islands on earth. It is a collection of vivid images of our striving to enrich the bounty and the beauty of this Eden. And it is the story of our growing sense of a paradise lost and the timeless struggle to discover origins and carry out the divine mandate—to name the works of the creation.

As a genre of publication, florilegia trace their origins to the close of the sixteenth century and according to Wilfrid Blunt, the authority on botanical illustration, reached their "highest popularity" during the seventeenth century. During that time, flowers were being more widely propagated for their beauty than for their utility. Technological advances in printing made it possible to produce works that took full advantage of the beauty of flowers and appealed to an ever-widening interest in horticulture and gardening.

At the hands of such well-known artists as Georg Dionysius Ehret (1708–1770) and Pierre-Joseph Redouté (1759–1840), flower painting reached new levels of refinement. Reproductions of their work are widely available and can be seen today decorating the walls of many homes and businesses. Their original published works are prized possessions in botanical libraries and holdings of private collectors, and when they become available on the open market they are all too frequently broken up and placed on the auction block, where they command high bids.

The best artists of florilegia set out to capture the gesture, or "signature," of the plant. This requires the highest degree of skill in painting technique. It also requires an intimacy with the subject that is achieved only by the most observant students of plants. And the best examples of botanical illustration in florilegia reach beyond the simple rendering of a plant to portray the essence of the subject: "Nature is visible, art concealed," according to Johann Wolfgang von Goethe, best known as author of *Faust* but also known to botanists for his seminal insights on the evolution of plants.

Mary Grierson's work as a botanical artist has been widely exhibited, featured in many books and other publications, and honored with five Gold Medals from the Royal Horticultural Society. From 1960 to 1972, she was Botanical Artist for the Royal Botanic Gardens, Kew. In *A Hawaiian Florilegium*, Miss Grierson is working in the time-honored tradition of Ehret and Redouté. In the rendering of the minute details we find her familiarity with her subjects, satisfying the stringent demands of the botanist for accuracy and at the same time elevating her subjects to works of art. Through her eyes and finely honed artistic skills, nature is framed to provide us with a cross-section of the world's most unique flora. Here a resonance can be found to satisfy our twin desires—to know plants in an intellectual sense while appreciating them at some deeper level, even as kindred spirits.

And how fortunate we are to have this visual resonance amplified by the writings of Mr. Peter Green, Deputy Director and Keeper of the Herbarium of the Royal Botanic Gardens, Kew, from 1976 to 1982, a member of the National Tropical Botanical Garden's Publications and Scientific Advisory Committees for eleven years, and a long-time student of Hawaiian plants and their lore. His descriptions are far ranging, revealing the scope of his own knowledge and his intimacy with the stories embedded in the names of this diverse assemblage of plants. In the *'iliahi* (sandalwood, *Santalum freycinetianum*) he traces the scent of money to deep roots where avarice and greed would in time change and obscure relationships of generations of people and of plants. *'Ōhi'a lehua* (*Metrosideros polymorpha*), the official tree of the island of Hawaii, tells a tale of our confusion in dealing with the varieties of form in nature. He takes us to a paradise where clandestine sex and seduction to aid in dispersal can lead to the *pāmakani* (*Viola chamissoniana*), a violet eight feet tall! And he offers us an invitation, indeed a plea, to become more familiar with the varied patterns in nature and challenges us to continue the stories of these plants.

In the section on "Endangered Endemics," Peter Green plots a trajectory from distant lands and distant times to new-found niches where new forms evolve. The frequent turn to woodiness is seen as a leitmotiv running through distantly related plants, while ever-tightening bonds of interdependence between birds, insects, and plants spur the creative process in new directions. Today these plants are on a path toward widespread habitat destruction, increased competition from alien species, population numbers reduced to a few individuals—and in all too many cases extinction. Once widespread, the attractive *uhiuhi* (*Caesalpinia kavaiensis*) is now known through fewer than fifty living individuals. In his description of the endemic Hawaiian genus *Cyanea*, Peter Green reminds us that twenty organisms in this alliance are now "lost forever." And several species of the native palm, or *loulu* (*Pritchardia*), are also known from only one or a few individuals in the wild and must be counted among the rarest palms in the world.

Through these essays we can sense the divine mandate to name as the first act in our stewardship of paradise. To name a plant is to assert some knowledge of its origins; it is speculation on relationships, a theory of history. And in many cases it is also an assertion of our desire to find immortality—"A Species of Eternity." Again and again we are confronted by confusing patterns, inadequate information, deception on the part of the plant, and self-deception on the part of the botanist. Nevertheless, we press on with this important work. As the twentieth century draws to a close, a new reality is pressing upward into our consciousness: the stories of these plants are our stories, and their struggle for survival is our struggle for survival.

The botanical garden, these treasured Edens, are where we gather the works of the creation in scientifically assembled and artful designs in celebration of our humanity. The National Tropical Botanical Garden, chartered by Congress in 1964, through its five gardens and three preserves in Hawaii and on the U.S. mainland, is pressing ahead on many fronts with the essential work of collecting, naming, nurturing, and sharing our love of plants. Our gardens are living laboratories, classrooms, and studios where each generation defines its place in nature and passes on what is learned to the next generation. In the timeless tradition of the florilegium, we are pleased to share this unique collection of *Botanical Portraits from Paradise*.

The struggle to bring this book to publication would appear to match the struggle of some of the plants depicted in it. It started back in the early 1970s when Mary Grierson came to Hawaii and was first introduced to the unique Hawaiian flora by Dr. William L. Theobald, director of the National Tropical Botanical Garden. Through the late 1970s and 1980s she returned to Hawaii six times to study, sketch, and paint. Dr. Theobald continued to support the effort right up to his retirement in 1993. The NTBG Publications Committee, so ably chaired by John Plews, Esq., was equally committed to the project, and through the generosity of the Edwin W. and Catherine M. Davis Foundation a wonderful idea was given form and substance.

Over the many years covered by the project, some of the paintings were dispersed by auction; whenever possible these were recovered to be photographed. Others were dispersed, never to be recovered. And a few of the paintings suffered the ravages of time spent in a tropical climate. With unflagging patience and incredible endurance, Mary Grierson and Peter Green maintained their interest in the project and their willingness to do whatever it took to see the work through to publication.

When I arrived on the scene of the NTBG in January 1994, the *Florilegium* was on hold. In John Thomas we were fortunate to find someone with the necessary editorial and diplomatic skills to shepherd the project through. It became a labor of love for John as it was for Dr. Theobald and for all the NTBG staff and Trustees who were associated with this project for almost twenty-five years. We can all be grateful for whatever it was that sustained them in their efforts to produce *A Hawaiian Florilegium*, a landmark publication that the National Tropical Botanical Garden is pleased to bring out with the assistance of the University of Hawaii Press.

William McK. Klein, Jr., Ph.D.
Executive Director
National Tropical Botanical Garden
July 1996

ACKNOWLEDGMENTS

Without the unfailing support of the National Tropical Botanical Garden, the paintings collected here would not have been possible. Many thanks are due to the NTBG staff for the help they provided me so freely and with such good will.

Friends outside the NTBG assisted me as well. The late John Allerton allowed me full use of his Guest House at Lawai-Kai—turning a room into a studio where I could work undisturbed—and gave me permission to collect specimens from his garden for painting. Liz McCoy offered kind hospitality when I first arrived; many of the paintings in this book were done while I stayed with Liz in Kalaheo. Donn Carlsmith invited me to spend time at his home in Keolahou, where I painted other subjects included in this book. Madeleine and Bob Emrick, Ralph and Catherine Dean, John and Steve Townsend, Sophie Cluff, Eleanor Crum, and many others provided friendship and help on my various visits to the Islands. I thank them all.

MARY GRIERSON
Richmond, Surrey

I wish to acknowledge the generous assistance and encouragement provided by Dr. William L. Theobald, Director of the Garden at the time of my visits and our initial inspiration for this book. In addition, I owe a debt of gratitude to the NTBG staff, especially botanists David Lorence, Tim Flynn, and Steve Perlman, for their help throughout the preparation of these materials. I also wish to thank Robert W. Read for advice relating to the palm genus *Pritchardia*, and George Staples, through correspondence, on the taxonomy of *Crinum*.

PETER S. GREEN
Royal Botanic Gardens, Kew

CONTENTS

INTRODUCTION TO THE FLORA OF HAWAII • 1

HAWAIIAN NATIVES

Iliau • *Wilkesia gymnoxiphium* • 6
Koa • *Acacia koa* • 8
Pua Maka Nui • *Hillebrandia sandwicensis* • 10
Pu'e • *Lobelia kauaensis* • 12
'Iliahi • *Santalum freycinetianum* • 14
'Ōhi'a Lehua • *Metrosideros polymorpha* • 16
Ōpuhe • *Urera glabra* • 18
'Ākala • *Rubus hawaiensis* • 20
Greensword • *Argyroxiphium grayanum* • 22
'A'ali'i • *Dodonaea viscosa* • 24
Hala Pepe • *Pleomele aurea* • 26
Pāmakani • *Viola chamissoniana* • 28
'Ohelo • *Vaccinium reticulatum* • 30
Naupaka Kuahiwi and Naupaka Kahakai • *Scaevola procera and S. taccada* • 32

ENDANGERED ENDEMICS

Uhiuhi • *Caesalpinia kavaiensis* • 36
Koki'o • *Kokia kauaiensis* • 38
'Āhinahina • *Argyroxiphium sandwicense* • 40
Munroidendron • *Munroidendron racemosum* • 42
'Ānaunau • *Lepidium serra* • 44
Hāhā • *Cyanea coriacea* • 46
Koki'o 'Ula'ula • *Hibiscus kokio* • 48
'Ōlulu • *Brighamia insignis* • 50
Loulu • *Pritchardia minor* • 52

POLYNESIAN COLONISTS

Hau • *Hibiscus tiliaceus* • 56
Taro • *Colocasia esculenta* • 58
Kukui • *Aleurites moluccana* • 60
Sweet Potato • *Ipomoea batatas* • 62
Breadfruit • *Artocarpus altilis* • 64

MODERN TRANSPLANTS

New Guinea Creeper • *Mucuna novoguineënsis* • 68
Devil Flower • *Tacca chantrierorum* • 70
Angel's Trumpet • *Brugmansia × candida* • 72
Bird of Paradise • *Strelitzia reginae* • 74
Night-Blooming Cereus • *Hylocereus undatus* • 76
Queen Emma Lily • *Crinum pedunculatum* • 78
Starfruit • *Averrhoa carambola* • 80
Shell Ginger • *Alpinia zerumbet* • 82
Calico Flower • *Aristolochia littoralis* • 84
Jack Fruit • *Artocarpus heterophyllus* • 86
Wood Rose • *Merremia tuberosa* • 88
Sacred Lotus • *Nelumbo nucifera* • 90
Banana Poka • *Passiflora mollissima* • 92
Liliko'i • *Passiflora edulis* • 94
Koster's Curse and Glorybush • *Clidemia hirta and Tibouchina urvilleana* • 96

Selected Readings • 99
Plant Index • 101

INTRODUCTION TO THE FLORA OF HAWAII

THE contemporary Hawaiian landscape is a drama of botanical paradox, irony, and variety. Few places anywhere are as caricatured in terms of their flora, yet the plants most identified with Hawaii are not natives. Visitors can see stunning ornamental favorites in any backyard as well in the leis they receive, but few of these species have been here more than a century. The same is true of pineapple, a symbol of Hawaii for much of the world. Even sugarcane, which has dominated the state's agricultural economy and much of its arable land, arrived here only with the Polynesian immigrants of the fourth or fifth century.

Behind this commonly perceived scene there is, indeed, a native Hawaiian flora—a flora remarkable to casual observers and botanists alike. Although they are hard to find and getting harder to find, there are perhaps a thousand different native species of flowering plant in Hawaii's 6,400 square miles—the great majority of them unique to the Islands. And among these species there is often enough variability of form to confound even the botanists whose business it is to classify them.

To tell a story that separates these distinct botanical strands and concludes with the present state of affairs in Hawaii—a dangerous state of affairs in which the ancient native plants are losing ground as well as visibility to the latecomers—it helps to begin at the beginning, before plants.

ORIGIN OF THE ISLANDS

The six well-known and often visited Hawaiian Islands, together with Niihau and Kahoolawe, are only the geologically most recent parts of the Hawaiian Archipelago, whose reefs and shoals stretch for 1,600 miles northwestward across the Pacific Ocean to Midway and Kure Islands. All the 122 islands, atolls, and shoals of the archipelago originated as volcanoes, above a mantle "hot spot" that has existed for over 70 million years. Eruptions above this hot spot piled up seamounts of pillow lava until they eventually burst through the surface of the sea; further eruptions then began to form shield volcanoes and sizable islands. Over this same period, the ocean floor has slowly, very very slowly, drifted westward, leaving this long string of volcanic islands, with the oldest in the northwest and the most recent in the east.

On a geological time scale, the Hawaiian Archipelago is still a dynamic system. Even now a new island, Loihi, is rising to the southeast of the island of Hawaii—"the Big Island." Its summit is now about 3,000 feet below sea level, but it will not become an island for many centuries. The Big Island itself is the youngest in the chain above the surface—less than one million years old—and is still dramatically active volcanically. Maui, which last saw an eruption in 1790, is slightly older and famous for its volcanic crater, Haleakala. Oahu's volcanoes are estimated to be 2 to 3 millions years old. Farther north and west, Kauai and Niihau are the oldest of the larger islands—estimated to be 5 to 6 million years old. Volcanic activity ceased there long ago, and erosion has reached an advanced stage, as is evident on Kauai at the famous Napali coast and Waimea Canyon. If the islands and shoals even farther west could be observed from below the sea, they would be seen to form a line of seamounts rising between 16,000 and 30,000 feet above the ocean floor, now eroded down to sea level and forming only low atolls.

THE ARRIVAL OF PLANTS

It is evident that the native plants and animals of Hawaii—by which we mean those that reached the Islands and evolved there long before humans could have introduced them—could not have migrated there over land. At one time this seemed to some people an unacceptable conclusion. The main Hawaiian Islands are about 2,500 miles from the nearest major landmass, the west coast of North America, and at least as far from other major Pacific island groups; how could such a variety of plants cross great distances of ocean by natural means? Notwithstanding the lack of geological evidence, some insisted that other Pacific islands, long vanished, must have once provided a land bridge between Hawaii and the sources of its flora. As might be suspected, better science proved the non-existence of such islands and provided more likely modes of transport—wind and ocean currents—and evidence that these have indeed been effective long-distance agents of plant dispersal.

Wind now seems the likely explanation for many Hawaiian plant species. The microscopic spores of ferns and very small seeds such as those of the *ʻōhiʻa lehua* can easily be lifted into the upper atmosphere by tropical storms. They can then be blown great distances before drifting down to sea level. An atmospheric jet stream passes over Southeast Asia at altitudes of 30,000–40,000 feet, where it reaches speeds exceeding 100 miles per hour before decelerating over the Pacific, near Hawaii. Seeds or spores from Southeast Asia blown up to these altitudes could have been carried to the vicinity of Hawaii in just a few days. Every now and then, some seeds or spores would land on an island, having dropped down by gravity or been washed out of the sky by high tropical rain clouds. Most would not survive such a journey in a fit state to germinate and grow, but with time some would.

Such jet stream transport is more than theory. Indeed, it has been calculated that 40–50 percent of the migrant ancestors of the present Hawaiian flora came from the Indo-Pacific region. A lesser proportion have links with plants of Central and South America; the famous Kona winds of winter, which come from the southeast, could have been the transporting agency for the ancestors of at least some of these plants.

Seeds could also have arrived by sea. Several strand plants, such as the beach naupaka, have seeds adapted to flotation in sea water and undoubtedly arrived in Hawaii this way. Still, the Islands are much less favored by ocean currents than by those of the air. The major Pacific currents completely pass by the Hawaiian Islands to the south and do not wash against their shores. This almost certainly accounts for the fact that many strand plants common on other Pacific islands, and adapted for seed dispersal by flotation, have nonetheless never reached Hawaii. Even the coconut, which could have floated to Hawaii on its own, was certainly introduced by Polynesian immigrants. It should be realized, however, that present members of the Hawaiian flora whose ancestors arrived by drifting may not be immediately recognizable. The seeds of the *wiliwili* (*Erythrina sandwicensis*), for example, do not float in sea water, but those of other species in this genus do. Perhaps an *Erythrina* arrived and, after establishing itself near the shore, produced offspring that slowly underwent modification and, in time, adapted to an inland environment such as that inhabited by the *wiliwili* today.

Seeds can also be transported great distances by migrating birds, or even by birds blown off course but fortunate enough to find an island on which to land. The seeds of marsh or bog plants can be imbedded in mud attached to a bird's foot or feathers. Other seeds are adapted for "hitchhiking" by means of barbs or hooks that catch onto feathers and are eventually shed when the bird preens itself at the end of a journey. Yet others are sticky, or become sticky when wet. Seeds of the genus *Lepidium* are like this, and ancestors of the *ʻānaunau* (*L. serra*) may have reached Hawaii that way.

DEVELOPMENT OF A UNIQUE FLORA

If Hawaii's geographic isolation seemed problematic for explanations of its flora, also mysterious was the great variety of species. Could so many distinct taxa arrive by air and water? More mysterious still would be the high proportion of species endemic to Hawaii, that is, species that do not grow naturally anywhere else in the world. According to the *Manual of the Flowering Plants of Hawai'i*, the percentage of Hawaiian endemics is particularly high: 89 percent of the flowering plant species and 65 percent of the ferns.*

The high degree of endemism and the number of species are less mysterious when one considers the vast time span involved. In the remote past, the ancestors of native Hawaiian plants arrived in the archipelago, spread to new, diverse, and evolving habitats, adapted further to their new microenvironments, and eventually changed into their present forms. It is estimated that all the Hawaiian native flowering plant species arose from fewer than 300 original immigrants. This multiplication of species, along with the high proportion of endemic species, illustrates what is technically known as adaptive radiation, whereby organisms adapt and establish themselves in various new habitats, change form to better survive in their new environments, and in the process become different species or even different genera. No other floristic region of the world exhibits adaptive radiation on such a scale.

Adaptive radiation is driven by changing and varied habitats, and Hawaii's geological history has made it an ideal provider. New lava flows, or new surfaces revealed by erosion and landslides, have been particularly characteristic of the ever-changing surface of the Islands. They have constantly provided new, open areas of land for occupancy—open areas not already dominated by other species. The fact that the Islands have arisen and then gone through the slow process of erosion sequentially has only added to the geological variability from one island to the next; recall, for example, that Kauai is several million years older than the Big Island, and that when Kauai was young there were older larger islands farther to the northwest. As an illustration of the related habitat variability, in spite of its tiny land area Hawaii is the only state with representative examples of all ten orders of soil recognized by the U. S. Soil Conservation Service.

Hawaii's climate could hardly be more favorable for plant life. The average temperature is mild and equable throughout the year, with generally moderate and constant humidity. But the weather is also variable. The prevailing tradewinds, hitting the mountains with their precipitous slopes and eroded valleys, produce areas of notoriously heavy rainfall. Other areas in the leeward rain shadow of the mountains are extremely dry. On Kauai, for example, the high top of Mt. Waialeale, often called the wettest place on earth, has an average annual rainfall of 450 inches, yet at Waimea, only about 16 miles away as the frigate bird flies, rainfall averages only about 25 inches a year. Temperatures vary too, for they drop about 3°F for every 1,000 feet one climbs. Today the Big Island reaches 14,000 feet, and Haleakala on Maui stands at 10,000 feet. Even on Kauai and Molokai the mountains reach 5,000 feet.

Thanks to this diverse weather, topography, and geology, the Hawaiian Islands exhibit a tremendous range of local climates. In combination, they have provided a multiplicity of new and different ecological niches—each a potential home for distinctive evolution toward new species. The vegetation of Hawaii has been classified into more than one hundred different stable plant communities: coastal, lowland, montane, subalpine, and alpine altitudinal zones are each potentially subdivided into dry, mesic, and wet communities, and within these are numerous unique ecologies dominated by a variety of different plants or plant types. No wonder there is such a diversity of endemic species. No wonder some of them display a particularly confusing variability.

HUMAN AND PLANT COLONISTS

It is no surprise that the equable climate and rich soils of Hawaii have encouraged the introduction of a great range of cultivated plants. The original Polynesian colonists, who began arriving some 1,800 years ago, carried with them two dozen or more kinds of plants most useful to them: banana, taro, breadfruit, sweet potato, yams, coconut, sugarcane, Indian mulberry, paper mulberry, *kukui*, *hau*, *'awa*. Some of these eventually escaped from their caretakers' cultivated plots and became

* Warren L. Wagner, Derral R. Herbst, and S. H. Sohmer, *Manual of the Flowering Plants of Hawai'i*, is the most current and comprehensive analysis of the native and naturalized flowering plants of Hawaii. For more information on this important source, on other works mentioned in this book, and on publications of related interest, see the Selected Readings list at the end of this book.

more natural parts of the landscape. Indeed, in this modern day several of these introduced plants are thought of as synonymous with Hawaii.

Though the Polynesian colonists added to the flora of Hawaii, they also diminished it. To cultivate these crops they cleared land—extensive areas of lowland forest, a type of plant community scarcely to be found today—and it must be assumed that some native species within those communities were lost. Still, this influence of humans on flora was minor compared to that yet to come—with European and American visitors in the late eighteenth century and then with colonists and later immigrants from all parts of the world.

In the sixty years following Captain James Cook's first visit to the Islands in 1778, seagoing merchants and new residents from America and Europe introduced more than a hundred new species, mostly fruit trees, vegetables, and ornamentals. Later in the eighteenth century, commercial planters introduced new agricultural crops such as pineapple, which along with sugarcane they developed into major agribusinesses. Acreage on a massive scale was cleared for these export crops and for cattle as well, with devastating losses to the native plant communities. At the same time, the East Asian and Southeast Asian laborers who migrated to Hawaii to work the cane and pineapple fields brough elements of their culture with them, including their own favorite food plants and ornamentals.

In the first half of the twentieth century, even the Hawaii Territorial Government contributed, planting hundreds of nonnative tree species in a misguided attempt to overcome the denuding effects of cattle and erosion. And the rest of the century has seen an explosion of exotic plant introductions by the horticulture industry. The overwhelming number were brought for ornamental purposes, and today many of them, like the bird of paradise or plumeria, are conspicuous parts of the lowland landscape—so much so that visitors and residents alike consider them to be native to Hawaii and are quite surprised to learn that they were actually introduced.

With all these new plants came animals, whose populations grew at outlandish rates. Apart from one bat, Hawaii had no native land mammals, so the coastal plains and mountains were open for invasion by pigs, rats, and many others. Dozens of ornamental taxa turned out to be invaders, too: the worst, banana poka, has smothered the native vegetation of hundreds of square miles. In combination, the clearing of land and the introduction of plants and animals have had a devastating effect on the native Hawaiian flora. Today, some four hundred Hawaiian plant species are considered to be either extinct or rare and endangered. Some have not been seen for years; others are reduced to a single known individual in the wild; yet others exist only in guarded cultivation. Many whole plant communities are vulnerable to interference by humans and the plants and animals they introduce. The high number of endemics in communities adapted to small microclimates makes the latters' vulnerability a particularly grave concern. The protection of individual species is undoubtedly important, but the safeguarding of entire unique communities on which the species depend is crucial. Since most of the endemics remaining are as yet unstudied, they represent resources of immeasurable potential value to us all. But they must be recognized and protected before they too disappear.

In the following pages, Mary Grierson has portrayed a selection of both native and introduced plants. Some were prized by the early Polynesians. Others are favorite ornamentals in tropical gardens worldwide. Still others illustrate subjects introduced for their beauty which have now become a menace, displacing vulnerable native species—a stark warning, almost too late, that plant introductions on oceanic islands must be strictly and rigorously monitored. And perhaps most important, this volume includes portraits of some of Hawaii's outstanding endemics. These may be thought to stand proxy for many others like them, vulnerable and threatened, about whose future an increasing number of thinking people are rightly becoming concerned. Many such people are actively involved in efforts to protect and increase the precious ecosystems that existed before the arrival of humans. We hope this book, with its beautiful images and descriptions of some very special tropical plants, will help preserve Hawaii's unique botanical heritage for the benefit and enjoyment of future generations.

Hawaiian Natives

ILIAU

Wilkesia gymnoxiphium

Family Asteraceae

WHEN in full flower, the *iliau* is perhaps the most bizarre of all the Hawaiian endemics. The elevated clusters of narrow leaves on the tops of tall, slim, more or less erect stems might look to the uninformed like palm trees. But when the plants flower this resemblance vanishes, for, shooting up to 15 feet, like fountains of yellow hanging daisies, appear whorl upon whorl of sticky flower heads. This plant is, of course, quite unrelated to the palms, and despite its narrow, grasslike leaves with lengthwise, parallel nerves it does not even belong to the group that contains grasses, lilies, and so forth but is a member of the daisy or sunflower family. Rarely in this species do the stems fork or branch, and then only when the growing point has been damaged, for example, by insect attack.

The genus *Wilkesia* is confined to Kauai and contains only two species, *W. gymnoxiphium* and the very rare and extremely local *W. hobdyi.* Both are found on the leeward mountains of the island. *Wilkesia gymnoxiphium* occurs in open sites on dry ridges or in dry to moist forest in the vicinity of Waimea Canyon, from about 1,500 to 3,500 feet altitude. In this area it is widespread but scattered, although it can occur in quite dense local populations. *Wilkesia hobdyi*, by contrast, probably exists in populations with a total of only about 350 individuals located on five dry ridges of extreme western Kauai, at about 900 to 1,400 feet.

Although so distinctive in its appearance, the *iliau* is nevertheless closely related to the well-known *'āhinahina*, or silversword (*Argyroxiphium sandwicense*, p. 40) and the greenswords (*A. virescens* and *A. grayanum*, p. 22), as well as to the several species in the much less spectacular genus *Dubautia*. Hybrids between the three species of *Argyroxiphium*, and between species of these three genera, have all been found in the wild or been synthesized in the laboratory. It seems that the genetic barriers to crossing one would expect to find between genera just do not exist in the case of *Wilkesia, Argyroxiphium*, and *Dubautia*, yet they are so distinct morphologically that recognition at generic rank is nevertheless appropriate—a striking example of the remarkable and unusual biology exhibited by the flora of Hawaii. In fact, despite their disparity in appearance, these three genera almost certainly arose many, many years ago, long before humans came to the Islands, from a single arrival of an immigrant tarweed from the western mainland of North America. All three genera are members of the same tribe, the Madiinae of the family Asteraceae (Compositae is the older, more traditional family name). The painstaking research that led to our present understanding of these plants was carried out by Gerald D. Carr and his colleagues.

The name *Wilkesia* commemorates Charles Wilkes (1798–1877), commander of the U.S. Exploring Expedition from 1838 to 1842. This expedition was sent out to discover more about the whales in the southern Pacific Ocean in the interests of the whaling industry, but Wilkes was instructed to carry out scientific observations in all branches of natural history. The strange-sounding epithet for this species comes from the Greek *gymnos* ("naked") and *xiphos* ("sword"), in allusion to the numerous swordlike leaves that form the rosettes at the top of the stems.

Wilkesia gymnoxiphium A. Gray
Plant growing on the western rim of Waimea Canyon, Kauai

KOA

Acacia koa

Family Fabaceae

THE best known of all the timbers in Hawaii is the *koa*, or *Acacia koa*, to give it its botanical name—the epithet having been derived from the Hawaiian name without alteration. This tree is found on the mountains of all the main islands except Kahoolawe and Niihau, generally between altitudes of 1,000 and 6,000 feet, and can in favorable circumstances grow to 70 feet or, rarely, even higher. The trunks of the largest trees in the high forest can have girths of 10 feet and reach up to 50 feet before they branch; in drier and more open places, however, branching takes place at a much lower level.

From these magnificent trees the early Hawaiians made their canoes, especially their war canoes and those used for long sea voyages. It is reported that in 1796 King Kamehameha the Great prepared a fleet of war canoes for the conquest of Oahu using *koa* from the forests behind Hilo on the Big Island. Over the years the big trees have been felled, and really large specimens are now rare. The timber, which is called "Hawaiian mahogany" because of its attractive appearance and color, is still used for making furniture and bowls when lumber of sufficient dimensions can be obtained, and smaller pieces are carved to manufacture novelties for tourists. The Hawaiians used *koa* wood for surfboards and canoe paddles as well as canoes, and timbers went into the construction of their huts. The bark too was used, to make a dye for staining tapa.

The flowers of *Acacia koa* are small and light yellow or cream colored, aggregated into fluffy heads made prominent by the numerous yellow stamens; the petals are quite small and insignificant. What appear to be the leaves are in fact highly modified leaf midribs, known botanically as phyllodes. True leaves can be seen on juvenile plants, where they are divided twice and bear numerous tiny leaflets, as in many other species throughout the genus *Acacia*. Occasionally these bipinnate leaves are also seen on the shoots that first appear after an injury to the tree. The curved, sickle-shaped, gray-green phyllodes are, however, the main photosynthetic organs, and because of them it is immediately possible to recognize *koa*. One or two other phyllode-bearing species of *Acacia* have been introduced to the Islands, but their phyllodes are straight or almost so, and their flower heads are solitary in the axils of the phyllodes, not in racemes or panicles. A related species, *A. simplex*, occurs on Vanuatu, New Caledonia, Fiji, Samoa, and Tonga, and a close species, *A. heterophylla*, is found on Reunion, one of the Mascarene Islands of the southwestern Indian Ocean. There are numerous phyllode-bearing members of this genus in Australia.

There is a certain amount of variability in *koa* from island to island, and separate if doubtful species have been recognized in the past. It is possible that three subspecies may be distinguished. Investigations by the late Professor J. P. M. Brenan, an authority on *Acacia* and Director of the Royal Botanic Gardens at Kew, were cut short by his untimely death, but this work is being continued by others.

Acacia koa A. Gray
Tree growing above 2,000 feet, toward Kokee, Kauai

PUA MAKA NUI

Hillebrandia sandwicensis

Family Begoniaceae

THE subject of this plate is one of the special plants of Hawaii. *Hillebrandia* is an endemic genus containing only one species, which occurs on four of the larger islands of the archipelago but not on the Big Island. *Hillebrandia sandwicensis* is generally found at altitudes above about 3,000 feet, in the damp and shady forest of ravines and gullies (although near the small museum at Kokee, on Kauai, it can be seen on a roadside bank). There are two Hawaiian names for this plant, *pua maka nui* (literally, "large-eyed flower") and *aka'aka'awa*.

Superficially, *Hillebrandia sandwicensis* looks much like members of genus *Begonia*, and in fact the two genera are closely related. The family Begoniaceae contains only the two genera, *Hillebrandia* (one species) and *Begonia* (with about nine hundred species)—that is, if one considers the rather doubtful genera *Symbegonia* (of New Guinea) and *Begoniella* and *Semibegoniella* (from South America) to really be part of *Begonia*. *Hillebrandia* differs from *Begonia* in two characters considered to be significant: an only partially inferior ovary, and distinct but very small petals, as well as sepals.

Hillebrandia was first introduced to cultivation in 1886, when seeds were sent to the Royal Botanic Gardens at Kew. It has, however, been reintroduced more than once since then and may be found in the collections of enthusiastic begonia growers everywhere. It grows from a tuber and can be treated like a begonia that likes fairly cool, if moist, conditions. It reaches a height of 3 or 4 feet, producing its delicate white and pink drooping panicles, which often appear terminal, from the axils of the upper leaves. If one looks up into the flowers, one may see the tuft of yellow stamens in the case of the male flowers, or in the females the curved stigmatic arms of the styles, which are at first yellow but turn a rich red as they age.

The genus name was given in honor of William Hillebrand (1821–1886), a German-born physician and botanist who lived in Hawaii from 1851 to 1871. In these two decades, Dr. Hillebrand left his signature on many facets of Hawaiian society. He was a founding physician of Queen's Hospital and the commissioner of the Bureau of Immigration under King Kamehameha V. Property first landscaped by him was eventually to become Honolulu's Foster Gardens. And he was author of the classic *Flora of the Hawaiian Islands*, published posthumously in 1888; it took more than a century for this Hawaiian standard to be superseded, by the *Manual of the Flowering Plants of Hawai'i* of W. L. Wagner, D. R. Herbst, and S. H. Sohmer.

Hillebrandia sandwicensis Oliv.
Plant collected at Kokee, Kauai

PU'E

Lobelia kauaensis

Family Campanulaceae

THE common garden lobelia, *Lobelia erinus*, has become naturalized in Hawaii, and this is the plant many people will think of when they hear the name "lobelia." But the native Hawaiian species of this genus are noteworthy and spectacular. They are all somewhat woody and frequently grow in wet, often boggy places in the mountains, far from any homes. One of them, *L. kauaensis*, is found in mountain bogs on Kauai, where it was referred to as *pu'e* by the early inhabitants.

Lobelia kauaensis develops a crown of leaves at the top of a stout, woody stem, growing gradually to a height of 3 to 6 feet, until it has built up sufficient food reserves to flower and fruit; once the seeds are ripe, the whole plant dies. The specimen that was the subject of this plate is in one respect atypical of the species: the inflorescence of *L. kauaensis* is normally branched, whereas in the two closely related species, *L. gaudichaudii* and *L. gloria-montis*, it is usually unbranched, as in this illustration.

Although most of the 350 or so species of *Lobelia* are herbaceous, it is remarkable that similar woody plants have developed on the mountains of the tropics in Africa, South America, and Hawaii. In Africa, where they can attain a height of 15 feet or more, they are often referred to as tree lobelias, but as in Hawaii their characteristic habitat is the bog. To accommodate these Hawaiian species, Otto and Isa Degener proposed segregating a genus called *Galeatella* in 1962 and then in 1965 distinguishing it from yet another genus, *Neowimmeria*. But species on a worldwide basis should be taken into account when making such a proposal. Considering the spectrum of variation in this group of plants, the separation of these genera solely to contain certain species from Hawaii, no matter how remarkable and spectacular, cannot be justified. Still, the distinctive morphological difference between the Degeners' two groups does suggest two different colonizations in Hawaii.

As with other members of the Campanulaceae in Hawaii, such as *Cyanea* (p. 46), the flowers are somewhat fleshy and curved. This character seems a natural adaptation to enhance pollination by species of honey creepers, native birds with a curved beak that can reach down to the nectar at the bottom of the floral tube. Some unrelated plants, such as *'ohe naupaka* (*Scaevola glabra*) and *koki'o* (*Kokia kauaiensis*, p. 38), also bear curved, tubular flowers adapted to these endemic and now too often rare or extinct birds.

The genus name *Lobelia* honors Mathias de L'Obel (1538–1616), a Flemish botanist who produced what was perhaps the first natural classification of plants. From 1584, L'Obel was botanist to King James I of England.

Lobelia kauaensis (A. Gray) A. Heller
Plant from Wahiawa Bog, Kauai

‘ILIAHI

Santalum freycinetianum

Family Santalaceae

ALTHOUGH commercially unimportant today, almost two centuries ago sandalwood began a short-lived but frantic surge through the economy and society of Hawaii. When trade was just beginning between Hawaii and the Western world—in the late eighteenth century—the fragrant wood of *Santalum* had long been prized in India, Burma, and especially China. So it is not surprising that, once they discovered that trees of this genus were to be found in quantity in Hawaii, enterprising ship captains soon opened up a lucrative trade.

Perhaps it is just a story, but it is said that sandalwood was discovered in Hawaii when the fragrance of some of the logs that had been taken on board for galley firewood was recognized by sailors who had traded this wood from other sources. The export of sandalwood from Hawaii probably began in the 1790s and was going well by 1810. Sandalwood became a monopoly of kings and high chiefs, who demanded logs of it in taxation. At first, collection was relatively easy, and so trade in this valuable export, the first from Hawaii, soon expanded. However, as the trees became more inaccessible, and as the chiefs' need for currency from outside grew, the trade in sandalwood became an intolerable burden on the people. Famine conditions began to develop as the taro fields and other crops were neglected while the search for sandalwood went ever higher into the forests and the logs had to be dragged ever farther to get them to the coast. The trade reached its peak between 1815 and 1827 and then declined rapidly in the early 1830s. Artifacts marking these times are the sandalwood pits to be found here and there, like the one on the road to Waikolu Lookout on Molokai. Pits were dug to match the size and shape of a ship's hold, so that when the pit was full another shipload was ready.

Although the genus contains about twenty-five species, distributed over the area of tropical Southeast Asia, Australasia, and the Pacific, there are four in the Hawaiian Islands, with five varieties. It seems probable that more species were once native to Hawaii. But extermination, particularly of the lowland populations, has left only part of the original representation on the Islands today. And what is left exhibits considerable, rather confusing variation. In the drier areas, for example, a *Santalum* may grow as a shrub up to about 3 feet high, but in moister wooded areas trees 30 feet high, and even 80 feet high, have been recorded. So it is no easy matter to arrive at a satisfactory classification of the relict native sandalwoods. Within the species *S. freycinetianum*, three somewhat overlapping varieties are recognized in the *Manual of the Flowering Plants of Hawai‘i*. The subject of this plate, var. *pyrularium*, has leaves that vary from oblong to narrowly elliptical and are slightly folded together lengthwise, small differences in the flower, and a rough-stoned fruit. It occurs only on Kauai.

The Hawaiian names for sandalwood are *‘iliahi* and *lā‘au ‘ala*, although the latter, which means "fragrant wood," can also apply to other scented timbers like the bastard sandalwood, *Myoporum sandwicense*. There appears to be no record that the Hawaiians burned the wood because of its fragrance, but they certainly knew of its properties because they used it in a powdered form to scent their tapa. The *‘iliahi* heartwood contains the most oil, and the older and darker in color it is, the stronger the scent and the more valuable the wood.

Santalum freycinetianum Gaudich. var. *pyrularium* (A. Gray) Stemmerm.
Plant from Waimea Canyon, Kauai

‘ŌHI‘A LEHUA

Metrosideros polymorpha

Family Myrtaceae

WITH its handsome scarlet flowers, many people consider *Metrosideros polymorpha*, variously called *‘ōhi‘a*, *‘ōhi‘a lehua*, or *lehua*, to be their favorite Hawaiian tree. In fact, these blossoms are the official flower of the island of Hawaii. Yet few species of tree anywhere exhibit greater variability, from dwarf, prostrate shrubs growing in bogs to tall forest trees over 100 feet high—the species is aptly named *polymorpha*. In part, this diversity may reflect the range of habitats in which the *‘ōhi‘a lehua* grows, but most of the variation is under genetic control, even though this may not yet have been fully investigated.

In this painting three variants have been depicted: on the right is the type variety, var. *polymorpha*, in the center is var. *glaberrima*, and on the left is a yellow-flowered form of var. *glaberrima*, called *lehua mamo*—the *mamo* being an extinct Hawaiian bird prized for its yellow feathers. The *Manual of the Flowering Plants of Hawai‘i* recognizes eight varieties and lists more than sixty names in synonymy, names that have been given for varieties and forms proposed in the past which are now generally considered unworthy of scientific recognition. The two varieties depicted in this plate are widespread, but others are restricted to single islands. Four other species of *Metrosideros* occur in Hawaii, all endemic.

The seeds of *‘ōhi‘a lehua* are small and easily carried by the wind; in times past they undoubtedly traveled long distances, colonized an island or island group, and then in isolation developed a new type unique to that vicinity. In fact, throughout the Pacific, on all the islands with significant elevation—from New Guinea, through the archipelagos of the Solomons and Vanuatu, to New Caledonia, Lord Howe Island, and New Zealand, and east to French Polynesia—there exists a group of closely related species. In the older literature, some of the Hawaiian species may be found under the name *Metrosideros collina*, which strictly speaking belongs to the representatives of this complex from Tahiti and the Marquesas, but there has been much debate and uncertainty in the past over the correct delimitation of these species.

The *‘ōhi‘a lehua* produces a fine, hard timber, as the genus name suggests, being derived from the Greek *metra* ("core" or "heartwood") and *sideros* ("iron"). As might be expected, it was much used by the original Polynesian islanders. Spears were made out of it, and straight timbers were used as rafters. Shorter pieces were made into beaters for tapa. In particular, *‘ōhi‘a lehua* wood was the medium for carving the images of gods. The special image, or *haku ‘ōhi‘a*, was carved from a single, carefully selected tree. Sacrifices were made before the tree was felled and then again, after the image had been carved, when the *haku ‘ōhi‘a* was finally erected on the *heiau*, a ritual temple. Wood from *‘ōhi‘a lehua* was also used in constructing the oracle pavilion and other structures of the *heiau*. In more recent times, it has been prized for furniture making.

The scarlet flowers are associated with Pele, the goddess of volcanoes, and so there are numerous legends connected with the *‘ōhi‘a lehua*. In the flowers, the conspicuous part is the tuft of bright red staminal stalks; the petals are rather insignificant. The flowers are held upright, and when fully open they hold nectar, which attracts the small, scarlet *‘i‘iwi*, one of the few, brightly colored endemic birds that still survive.

Metrosideros polymorpha Gaudich.
Plants collected on the slopes of Mauna Kea on the island of Hawaii

ŌPUHE

Urera glabra

Family Urticaceae

IT cannot be said that the flowers of this member of the stinging nettle family are particularly attractive. Throughout the whole family, the flowers are small and rather insignificant, and apart from some members of the genus *Pilea* with colored and otherwise attractive leaves, none of them has horticultural merit. Nevertheless, the family is well represented in the Hawaiian flora, with seven genera. Two of these, *Neraudia* (with four species) and *Touchardia* (one species), are to be found only on these islands. Altogether, fourteen species are endemic, although some authorities would recognize more. *Urera*, the genus to which the subject of this plate belongs, comprises about thirty-five species distributed throughout Africa and Madagascar, Hawaii, and tropical America. They occur as separate male and female trees, and the branch that has been painted here came from a female.

Urera glabra, which in some books may be found under the incorrect name *U. sandwicensis*, occurs on all the main islands of Hawaii except Niihau and Kahoolawe. This species is somewhat variable in stature and in characters of the leaf, such as shape, hairiness, and the length of the stalk. The *Manual of the Flowering Plants of Hawai'i*, although treating them all as part of *U. glabra*, suggests that the variants might be recognized as four subspecies, and certainly the plant from Kauai illustrated here, with its diffuse inflorescence, long leaf stalks, and ovate leaves with five or six main side veins, was in fact recognized as a separate variety, var. *kauaiensis*, by Joseph Rock in his classic *Indigenous Trees of the Hawaiian Islands*.

Rock recorded the Hawaiian name *hona* for this plant, but this was probably a name confined to Kauai, for the usual name to be found in the literature is *ōpuhe* or *hōpue*. The strong and pliable fibers in *ōpuhe* bark were used by early Hawaiians to make a fine cordage that was principally made into fishing nets, although it was also woven into the network on which feathers were attached in the famous feather capes, cloaks, and helmet covers. Similar nets were used as the basis for raincoats made of the leaves of ti (*ki*, *Cordyline fruticosa*), and others, with a more open network, were employed to carry bowls and other rounded containers. The fibers of *ōpuhe* are very similar, and may be a little inferior, to those of the closely related *olonā* (*Touchardia latifolia*), also a member of the stinging nettle family which was used for these same purposes.

The name *Urera* comes from the Latin *ure* ("burn" or "sting"), alluding to those members of the genus that bear stinging hairs like those in the genus of stinging nettles, *Urtica*. Thankfully, such hairs do not occur on the Hawaiian species. The epithet *glabra* means "without hairs," despite the fact that some of the variants of this species do in fact have hairs on their leaves.

Urera glabra (Hook. & Arn.) Wedd.
Plant from the Awaawapuhi Stream, Kokee, Kauai

ʻĀKALA

Rubus hawaiensis

Family Rosaceae

CONSIDERED as whole, there are few more problematic genera than *Rubus*, which contains the raspberry and the blackberry. Estimates of the total number of species vary from about 250 to well over 3,000. The higher number is a reflection of what are sometimes called "microspecies," in which the normal genetic processes of reproduction have been circumvented. Seeds from these plants generally reproduce the parental type exactly, just as though they were clonal. But infrequent minor mutations can occur, or occasional hybridization between these entities can produce slightly different individuals, and these then reproduce themselves exactly. In this way new, uniform populations that are consistently different can arise. Such entities, when recognized as distinct, described, and given a name, are microspecies. They are not, however, comparable biologically with normal out-breeding species.

There are two Hawaiian endemics in this genus, which fortunately do not seem to vary like some of those on the mainland. One, *Rubus macraei*, is a rare, rather prostrate plant confined to East Maui and the Big Island. The other, *R. hawaiensis*, is more widespread, found in moist woodland and forest between 2,000 and 9,000 feet on all the bigger islands except Oahu. It may bear prickles or be without them (the so-called var. *inermis*). The pink, wavy petals are not very showy, but the large, dark red, juicy fruits stand out and are quite attractive. They are edible and were eaten by the early Hawaiians, in spite of being tart and somewhat bitter. To our modern tastes they are, perhaps, better made into pies or preserves. Attempts have been made to cross this species with others known to produce tasty fruit, but without much success.

As well as eating its fruit, the early Hawaiians used this plant to obtain a pink dye for their tapa, hence the native name *ʻākala,* which means "pink," and sometimes *ʻākalakala* or even *kala*. It has also been said that they used the bark of *ʻākala* to make an inferior type of tapa, but doubt has been cast on this suggestion, and when one thinks of the size of the stems in *Rubus* it does seem unlikely.

As well as the two native species, five others have been introduced and become naturalized in Hawaii. All five are potentially dangerous to the native vegetation, and two of them are already serious problems. One of these, *Rubus rosifolius*, sometimes called thimbleberry or Mauritius raspberry (although a native of Asia introduced to Hawaii by way of Jamaica), has leaves with five or seven leaflets, rather small prickles, white flowers, and fruit that is bright red, an inch or more long, thimble shaped, and somewhat insipid in taste. The other, an even more serious weed, is the so-called prickly Florida blackberry, *R. argutus*, sometimes called *R. penetrans*. This pest is actively invading and smothering anything it can cover on Kauai, Oahu, Maui, and Hawaii, replacing the native undergrowth with impenetrable thickets of arching branches and sharp, recurved thorns. Its large white flowers can appear attractive in the spring, at least to those unaware of the damage it is doing, and its black, juicy fruit has a pleasant flavor. Although originally brought to Hawaii with the best of intentions, like a few other noxious weeds this *Rubus* species is an object lesson in the dangers of introducing mainland plants to the vulnerable and delicately balanced floras of oceanic islands.

Rubus hawaiensis A. Gray
Plant on the southeast slope of Mauna Kea, Hawaii

GREENSWORD

Argyroxiphium grayanum

Family Asteraceae

THE genus *Argyroxiphium* contains only five species. Of these, the *'āhinahina*, or silversword (*A. sandwicense*, p. 40), is the best known. The other members, including the greensword depicted here, are less often heard of, presumably because they are less in the public eye. The greensword grows in both East and West Maui. This is the only island where it is found, and it is restricted there to altitudes between about 4,000 and 6,500 feet, mostly in or near the margins of bogs, in areas of mountain mists—a very different habitat from that of the popular silversword.

The name greensword, which has been applied to both *A. grayanum* and another species, *A. virescens*, reflects the fact that the leaves of the two are not covered with the silky, silver hairs of the silversword. The former two species are closely related, and there has sometimes been confusion between them, with the subject of this plate occasionally being referred to as *A. virescens* var. *paludosa*, a name that correctly should apply only to the naturally occurring hybrid between the two. In conservation terms, however, *A. virescens* is classified as "presumably extinct," for it has not been seen for fifty years. *Argyroxiphium grayanum* has also been recorded as hybridizing with no less than three species of the related genus *Dubautia*. In fact, these two genera, together with *Wilkesia* (p. 6), have all been crossed successfully in the laboratory. This extremely close and common genetic relationship has been attributed to a possible single ancestor that arrived thousands of millennia ago from western North America, an ancestor that today would probably be recognized as one of the tarweeds. Such an immigrant could well have arrived on an island that once existed to the west of Kauai before the elements wore it down almost to sea level, like the present-day Nihoa Island or Necker Island—and before the Hawaiian Islands as we know them had even appeared above the ocean. Then, as it migrated eastward onto newer and newer islands, its descendants would have occupied different available habitats, diversified, and evolved into the related genera and species we see today.

As with the silversword and *iliau*, an individual greensword plant usually dies after flowering. The plant vegetates for several years, building up food reserves. At the right stage, when mature, the growing point develops into an immense inflorescence that shoots up as much as 3 feet (and 8 feet in the case of the silversword). After the flowers have withered and the seed set, the whole plant dies. Not only has it used up its reserves but, perhaps more important, the usually single growing point has been subsumed in the development of the inflorescence. Sometimes a side shoot develops while the plant is growing. When this happens, only the shoot that flowers dies, and the side shoot in turn grows on to develop its own inflorescence.

This species was first collected in 1870 on the slopes beneath Eke Crater, West Maui, by William Hillebrand and described by him in the first published flora of Hawaii, his *Flora of the Hawaiian Islands*. Hillebrand described the plant as a species of *Wilkesia*, but later it was rightly reclassified in *Argyroxiphium*. The specific epithet, *grayanum*, honors the famous nineteenth-century American botanist Asa Gray (1810–1888), a Harvard professor, correspondent and friend of Charles Darwin, and the leading American protagonist of Darwin's theory of evolution.

Argyroxiphium grayanum (Hillebr.) O. Deg.
Plant collected on the upper wet slopes of Haleakala, East Maui

ʻAʻALIʻI

Dodonaea viscosa

Family Sapindaceae

SOME plant species are scarcely variable at all, even in the wild—take the *ʻāhinahina* (p. 40) or *iliau* (p. 6), for example. Others exhibit great differences in such characters as stature, color, leaf shape, and hairiness, and the *ʻaʻaliʻi* is one of them. In fact, E. E. Sherff, sometimes in collaboration with Otto Degener, described and named no less than twenty-five varieties and ten forms of what in the *Manual of the Flowering Plants of Hawaiʻi* is treated as a single species without varieties or forms, *Dodonaea viscosa*. This plant may also be found listed under the name *D. eriocarpa*, and the Hawaiian names include *ʻaʻaliʻi*, *ʻaʻaliʻi kū ma kua*, and *kū makani*—the last meaning "wind-resisting," one of the attributes of this plant. To write their account for the *Manual*, the authors analyzed no less than five hundred specimens collected throughout Hawaii. The characters of leaf size and shape, the size and hairiness of the fruiting capsule and its number of wings and degree of inflation—the characters on which the numerous earlier varieties and forms had been based—were studied, but no clear, discrete entities could be recognized. The variation seems totally haphazard. Individual plants are different from one another, and seem distinct, but the differences are not significant when it comes to recognizing reliable species or varieties, any more than are the color of eyes or hair, or the shape of the nose, in the species *Homo sapiens*.

Perhaps the most striking feature of this plant is its attractive fruits, with their diversity of color. The plate illustrates three typical shades, but there are also intermediates, and some are even darker than that depicted, almost black. There is also variation in the hairiness of the fruits and the degree to which they are clustered. The flowers themselves lack petals and are insignificant, but within the species they show considerable diversity in sexual expression. Some individuals bear flowers with both stamens and ovary; others have flowers that are either male or female.

The *ʻaʻaliʻi* is widespread throughout the Hawaiian Islands, especially in the drier areas, and it is likely that even more of it existed before the natural vegetation at lower altitudes was cleared to make way for sugarcane and other crops. The wood is extremely hard and durable and was much used by the early Hawaiians. The trees are relatively small, usually no more than 25 feet tall, but stems from the largest of them were used to make posts for the construction of houses. Straight stems were made into light spears for fishing, notably for squid. The hard, heavy wood, which sinks in water, was also used to make bait sticks. These were cylindrical in shape, but gradually tapered to a narrow end notched to hold a line, and were about 2 feet long and two or three inches in diameter. Before use they were smoothed, charred over a fire, and finally rubbed with *kukui* oil or coconut oil. The bait stick was lowered over the side of a canoe to a depth of 20 feet or so, a few feet above the sea floor where it could still be seen through the clear water. When a fish was attracted to it, the bait stick was gradually drawn toward a net, in which the fish was then caught.

Quite a different use of the *ʻaʻaliʻi* was, and still is, the weaving of its attractive and colorful fruits into leis. The red fruits have also been used to make a dye for decorating tapa.

Dodonaea viscosa Jacq.
Plant collected from the Iliau Loop Trail, Waimea Canyon, Kauai

HALA PEPE

Pleomele aurea

Family Agavaceae

WHEN naming and classifying plants, botanists often see similarities between different species and so group them into one genus. But they may more easily notice the differences that exist between the species. As a consequence, there are botanists who lump species together because they see similarities and others who keep them apart because they see differences—what are jokingly referred to as "lumpers" and "splitters." It may be that psychological makeup determines to which class a botanist belongs, but ideally they should all note both the similarities and the differences and judge the significance of each. Wild plants are not man-made though, and, due perhaps to past hybridization or to incomplete differentiation over time between one group of species and another, the differences between such groups may be such that some will think them relatively minor while others consider them adequate enough to constitute what they would call a genus. When this happens, the recognition or rejection of a suggested genus depends in great part on the experience, standing, and authority of the botanist who makes the decision. One's judgment can also depend on whether the particular group has been studied locally or looked at on a worldwide scale, for what can seem discrete and distinct groups locally can merge when the constituent species are studied throughout their whole range.

The Hawaiian members of the group represented by this plate are a case in point, for the name depends on which authority is followed. Although many botanists would tend to recognize a large and variable genus *Dracaena*, the recent and standard works on the Hawaiian flora maintain the segregate genus *Pleomele*. As entities, they differ in whether the six perianth segments, or "petals," are joined in a tube or free from each other, and whether the filaments of the stamens are slender and not enlarged. Small differences, you may think, but when taken in the context of all the other related genera, and on a global scale, such floral characters become significant. Whether you call this plant *Dracaena aurea* or *Pleomele aurea*, however, there is no doubt that it is a handsome and beautiful subject. Perhaps, to remove ambiguity, it is easier to use the old Hawaiian name, *hala pepe*, but there is also the Western common name that has been coined for it, the golden dracaena.

Hala pepe is endemic to the island of Kauai, and five other *Pleomele* species are found in different parts of the archipelago, all of them native to Hawaii. *Pleomele aurea*, which can grow into a tree 15–25 feet tall, is perhaps the most attractive of them all. It grows in the rough woodland around Kokee at a considerable altitude, and elsewhere on Kauai in diverse mesic forest down to 400 feet. The golden trusses of pendant flowers are followed by globose berries that turn dark red when ripe and contain three rounded seeds, unless one or two of them fail to develop. It is interesting to note that the Hawaiian species of *Pleomele* grow well from seed but are more difficult to propagate from cuttings, unlike many members of the genus *Dracaena*.

The name *Pleomele* is derived from the Greek *pleios* ("many") and *melon* ("apple"), although the many berries produced are scarcely apple-like! *Aurea* is Latin for "golden" and refers to the color of the flowers.

Pleomele aurea (H. Mann) N. E. Br.
Plant growing at Kokee, Kauai

PĀMAKANI

Viola chamissoniana

Family Violaceae

VIOLETS have long been a favorite and romantic flower, but together with pansies they are only part of the botanical genus *Viola*, which contains about five hundred species. Most of them come from the north temperate regions, but several are native to the Andes. Seven species are now recognized from Hawaii, and comparative studies point to the possibility that these species arose from forebears that arrived by way of long-distance, chance dispersal from South America. The remarkable thing about the Hawaiian representatives of this usually herbaceous group is that six of the seven are shrubby with woody stems. *Viola chamissoniana* subsp. *tracheliifolia*, the subject of this plate, can grow to as much as 8 feet tall.

Before the revisionary work that went into the preparation of the *Manual of the Flowering Plants of Hawai'i*, this taxon had been treated as an independent species, *Viola tracheliifolia*, but it appears that it is not sufficiently well differentiated and distinct from *V. chamissoniana*—hence its relegation to the lesser taxonomic rank of subspecies. Subspecies *tracheliifolia* occurs on Kauai, Oahu, Molokai, and Maui, but not on the Big Island. The rare subspecies *chamissoniana* is found on the dry cliffs of western Oahu. A third subspecies, *robusta*, grows in Molokai's wet forests.

Despite its woody character, the flowers are shaped like those of a typical violet. They are white, tinged with purple, especially around the spur, which contains nectar that attracts pollinating insects. The leaves too are similar to those of many other violets (although in the botanical epithet *tracheliifolia* they are likened to those of the quite unrelated Mediterranean genus *Trachelium*). After the flowers come the typical three-cornered capsules which, when they mature and dry out, burst open violently and scatter their seeds because of the tensions set up in the capsule walls. The seeds of many *Viola* species bear at one end an oily body, called an elaiosome, which is attractive to ants, and this ensures their further dispersal. It is recorded in this species, as in many and perhaps most members of the genus, that small, incompletely developed flowers with diminutive petals are produced toward the end of the flowering season and pollinate themselves (technically called cleistogamous flowers). These produce seed even if the normal flowers have not been fertilized.

Viola was the Latin name for many sweet-smelling flowers, including the sweet violet, and was derived from *wion*, a Greek term for the color violet. The epithet *chamissoniana* commemorates Ludolf Adelbert von Chamisso (1781–1838), a poet, author, botanist, and explorer. As a member of Lt. Otto Kotzebue's 1815–19 voyage of exploration on the Russian imperial naval ship *Rurich*, Chamisso discovered and for the first time collected many plants on the island of Oahu in 1816 and 1817.

Viola chamissoniana Ging. subsp. *tracheliifolia* (Ging.) W. L. Wagner, D. R. Herbst & Sohmer
Plant collected on the west flank of Waimea Canyon, Kauai

‘OHELO

Vaccinium reticulatum

Family Ericaceae

ONE of three species of genus *Vaccinium* native to Hawaii, the *‘ohelo* ranks among the Islands’ most popular wild plants. On Kauai, Oahu, and Molokai it is rather rare, but it is common in the subalpine shrublands of Maui and Hawaii between about 2,000 and 12,000 feet, reflecting its character as a low, bushy pioneer of open and disturbed habitats like lava flows and cinder beds. It is especially common in the vicinity of the Kilauea Caldera. This association with the outpourings of volcanoes, combined with its bright red flowers and berries and with young foliage frequently tinged with red, seems a natural enough explanation for the original Hawaiians dedicating this plant to Pele, the fire goddess of volcanoes.

In season, the very palatable berries of the *‘ohelo* can be abundant, but it is said that before they may be eaten an *‘ohelo* branch must be cast into the firepot of Kilauea to placate the fearsome Pele. Carl Skottsberg, a Swedish botanist who worked on the flora of Hawaii between the two world wars, gave the name *Vaccinium peleanum*, in honor of the goddess, to a variant of this species which occurs on volcanic Mauna Loa. Now, however, this and seven other so-called species which had earlier been described from Hawaii are not generally thought sufficiently distinct for recognition as separate species.

Berries of this genus are much prized. In English-speaking countries, various species bear blueberries, huckleberries, bilberries, whortleberries, or blaeberries, the name depending on where they are found. Even the less closely related cranberry is a *Vaccinium*. In all, it is estimated that there are about 450 species worldwide, mostly from temperate climates, and all are found in the Northern Hemisphere except for those occurring on high mountains in the more tropical regions. The species closest to those native to Hawaii are found on islands in the south and southwest Pacific, especially *V. cereum* of Tahiti.

Flowers and berries are borne on the *‘ohelo* throughout most of the year, but the peak of berry production occurs between June and September. Commonly made into jams or jellies, the berries can also be eaten raw. Their characteristic color is bright red, but they may be reddish-purple, dark bluish-purple, or at the other end of the color range even yellow. They are often covered with a waxy bloom. The seeds are small but are produced in great number. It is reported that *‘ohelo* berries were the main source of food for the *nene*, or Hawaiian goose, once near extinction but now protected and perhaps recovering. The flowers too are characteristically red, but individuals with yellow or greenish-yellow flowers can be found.

Close inspection of *‘ohelo* stamens shows that they liberate the pollen grains through two pores at one end of the anthers. In most plants, by contrast, anthers release their pollen by splitting down their sides, which then curl back and open wide. Compared to this typical arrangement, one might think that an opening through a small pore must be less than efficient. But at the opposite end of the *‘ohelo* anthers are two appendages, or “horns,” which rest against the inside of the corolla—the urn-shaped, united petals. When an insect visits the flower to get nectar, it has to approach from below. In landing, or poking within, it inevitably vibrates the corolla. This in turn agitates the anthers and shakes the dustlike pollen onto the insect hanging below.

Vaccinium reticulatum Sm.
Plant growing near Kilauea Crater on the island of Hawaii

NAUPAKA KUAHIWI AND NAUPAKA KAHAKAI

Scaevola procera and *Scaevola taccada*

Family Goodeniaceae

THE upper plant portrait in this plate is of the mountain naupaka, *naupaka kuahiwi.* The lower is that of the beach naupaka, *naupaka kahakai.* Although they are both native to Hawaii, they occur in quite different habitats, as may be inferred from their names. The beach naupaka, being a strand plant, occurs throughout the islands of the Pacific and Indian Oceans, but the mountain naupaka is endemic to the Hawaiian Islands and can be found only on the islands of Kauai and Molokai, in moist forest, usually between about 2,000 and 4,000 feet altitude.

Mountain naupaka, *Scaevola procera,* grows as a shrub or small tree up to 10 feet tall. It is quite complex in form, and some authorities suggest that it may have arisen by hybridization between two other complex and variable species, *S. gaudichaudiana* and *S. mollis*. Further scientific investigation is called for, but there is no doubt that hybridization has played its part in the development of Hawaii's remarkable flora.

Beach naupaka, *Scaevola taccada,* is a widespread species, occurring on seashores from East Africa across the Indian Ocean, through the Southeast Asian archipelago, on tropical Australia, and across the Pacific within the tropical and subtropical belts. It varies in the degree of hairiness, and the leaves are often quite silky or, to use the technical term, sericeous. This is the basis of the synonym that can be found in some of the literature, *S. sericea.* The nomenclatural complexities involved in adjudicating between these two competing names almost defy belief, but, after all the arguments, *S. taccada* appears to be the correct one.

Like the seeds of most if not all strand plants, those of the beach naupaka float and are not killed by sea water. Clearly this is the way it has spread across the two oceans, and also how it reached Hawaii. The fruits are white, somewhat globose, and about half an inch in diameter—the origin of what is said to have been a rarely used Hawaiian name for this species, *huahekili,* the word for hailstones. With its handsome light green leaves, sturdy growth, and ability to thrive within the range of the ocean's salt water spray, it is a most useful plant for landscaping in shoreline gardens. It is also relatively easy to propagate by seed or cuttings.

The unusual shape of the flowers in this group of plants, appearing to be only half a flower, and the lower half at that, has led to romantic myths in Hawaii. According to one, perhaps of relatively recent origin, because of a quarrel with her lover, a maiden tore a naupaka flower in two and then declared that she would not love her old sweetheart again until he brought her a perfect flower. He searched all over the Islands—but in vain, for these flowers, whether they grew on the seashore, on the plains, or in the mountains, had become but half-flowers. He died of a broken heart.

The other myth also concerns two lovers, who in this case enraged Pele, the goddess of volcanoes. She pursued one of the lovers into the mountains, but the gods took pity on him and transformed him into the mountain naupaka. The other lover she overtook at the edge of the ocean, but the gods turned her into the beach naupaka. From that time, the lovers have remained separated; the half-flowers of the boy still bloom alone in the mountains, and the half-flowers of the maiden blossom alone on the beach.

The flower shape is also said to have inspired the selection of the genus name, *Scaevola,* this time through an ancient European legend. As it

Scaevola procera Hillebr. and *Scaevola taccada* (Gaertn.) Roxb.
Plants collected at 4,000 feet altitude near Kokee and near the shoreline, Kauai

is told, when Lars Porsena, king of Clusium in Etruria, was besieging Rome in 507 B.C., Gaius Mucius took it upon himself to assassinate him. Mucius managed to gain entry to the royal tent but, not knowing the king personally, in error killed Porsena's more sumptuously dressed secretary instead. Mucius was seized and brought before the royal court of justice, where he claimed that he was one of three hundred noblemen who had sworn to kill the king and had been chosen by lot to be the first to attempt the assassination. When threatened with torture, he thrust his right hand into the fire on the altar and held it there until it was destroyed, to show how he despised pain. Porsena, in admiration of his courage, freed Mucius and, in fear of the valor of the Romans, abandoned the siege. In appreciation, the Roman state made Mucius a grant of land, and from then on, in remembrance of his feat, he was called Scaevola, "the left-handed one." Centuries later, when Linnaeus came to name this genus of plants, he noted that the flower petals resembled a hand spread out, so he commemorated Gaius Mucius Scaevola. And the flowers, which start out white, turn brown with age as though scorched by fire.

Endangered Endemics

UHIUHI

Caesalpinia kavaiensis

Family Fabaceae

ALTHOUGH *uhiuhi* was formerly more widespread, and today is still known from four islands—Hawaii, Kauai, Oahu, and Lanai—it is said that fewer than fifty individuals of this attractive species now exist. As a consequence, it has been listed as "Endangered" under the U.S. Endangered Species Act. It grows in dry forest, much of which has been destroyed and converted to grassland in the Islands. Other, small populations—particularly in North Kona, on the Big Island—grow in a kind of savanna on particularly rugged *a'a* lava covered by fountain grass, *Pennisetum setaceum*, to breast or even shoulder height. This terrain is less likely to be disturbed by man, but the *uhiuhi* there has suffered significant losses in wildfires fueled by this alien fire-adapted grass.

In many of the works on Hawaiian plants, this species is referred to as *Mezoneuron kavaiensis*, but it is now generally accepted that the genus *Mezoneuron* should be classified with *Caesalpinia*. The only real difference between them appears to lie in the pod, where in the former it is winged along one edge but in the latter wingless. Not only does this seem an insufficient basis on which to separate two genera, but some species that have always been considered to belong to *Caesalpinia* bear a distinct, if narrow, wing on their pods. There are three other species of *Caesalpinia* in Hawaii. Two of them, *C. bonduc* and *C. major*, may be indigenous. The third, *C. decapetala*, was certainly introduced. The most nearly related species to *C. kavaiensis* appears to be one that is native in New Guinea and Australia, *C. scortechinii*.

Early Hawaiians used the wood of *uhiuhi* in a variety of ways. It is particularly hard, and this is the quality that was put to use. For instance, it was made into digging sticks called ō'ō, used when planting taro or other crops; evidently some of these tools were shaped into flat blades, but others were pointed. Tools for cutting off the top part of taro corms were also made of this wood. The corm, which was being harvested, was separated from a portion that still had the leaf bases attached; this could then be planted for the next crop. The wood was used to make tapa beaters, and in the construction of the most important houses it was used for the corner posts. For fishing it was fashioned into shark hooks and, when long straight sticks of it could be obtained, they were made into fishing spears, 6–7 feet long or more, and were used to spear squid as well as fish. Weapons of war were made of it, spears and daggers as well as clubs—the last also being used in fishing to stun and kill *'ahi* and other large fish. But perhaps the most interesting use of this hard and durable wood was in making the runners for sleds (*hōlua*), which were used in an aristocratic sport. Special sloping tracks were made on the side of a hill, with rocks as foundation, covered with earth beaten down hard to give a smooth sloping surface, and then covered with dry grass so that the sled would run easily over it. These sleds had runners 12 or more feet long made of *uhiuhi* wood, and competitions were held to see who could travel the farthest down the slope and onto the level ground at the bottom.

Caesalpinia kavaiensis H. Mann
Plant growing at the National Tropical Botanical Garden, Lawai, Kauai

KOKI'O

Kokia kauaiensis

Family Malvaceae

ONE of the plant groups that has diversified most in Hawaii is the mallow family, which contains not only the ever-popular *Hibiscus* but also two remarkable endemic genera, *Kokia* and *Hibiscadelphus*, whose representatives are all now either extinct or endangered. In both genera the flowers appear to be adapted for pollination by birds with a curved beak, such as the endemic Hawaiian honey creepers. Many of these birds, like the plants whose nectar was their source of food, are now extinct. Without their associated pollinators, with forest destruction and its related competition from woody weeds from the continents, and probably with the consumption by rats of much of any seed that might be produced, it is no wonder these endemic and rather specialized plants have disappeared or hang on today in small numbers, perhaps before vanishing forever.

It is generally recognized that there are, or were, four species of *Kokia*. One of these, *K. lanceolata* of southeastern Oahu, has been extinct for almost a century. Two are federally listed as endangered: *K. cookei*, now known only in cultivation, and *K. drynarioides*. The subject of this plate, *K. kauaiensis*, although surviving in a few small populations in the mountain valleys of western Kauai, is classified as a candidate for immediate listing as endangered under the U.S. Endangered Species Act and as "Vulnerable" by the International Union for the Conservation of Nature and Natural Resources.

Plants of *Kokia* grow into beautiful trees, well worth cultivating. The flowers are striking, and even in bud, or after the brick-red petals have dropped, they are marked by three large, broad and conspicuous bracts. These persist in the fruiting stage and are strongly tinged with red, the almost intrinsic color of *Kokia*; even the leaves have red veins and are often suffused with red. It is noteworthy that Joseph Rock, in his *Indigenous Trees of the Hawaiian Islands*, coined the name "native red cotton" for this species. The calyx consists of an urn-shaped cup, a little under an inch deep, which is closed over the rest of the flower as it develops. The fruit consists of an ovoid, woody capsule that splits open to reveal one to five seeds. These are densely covered with reddish-brown hairs up to nearly a half-inch long. It has been suggested that these hairs may have enabled seeds of the ancestor of *Kokia*, the original migrant to the Hawaiian Islands, to float across the ocean. Certainly, the seeds of present-day *Kokia* species will float for several days, though as soon as the hairs become thoroughly wet they sink. Still, the possibility of the original progenitor arriving by sea must not be dismissed lightly. The most closely related genus to *Kokia* appears to be *Cephalohibiscus*, which has one species only, in New Guinea.

That *Kokia* was less rare in former times is demonstrated by the fact that the early Hawaiians, who called it *koki'o*, used its bark as a source of red dye with which they both colored and preserved their fishing nets. The generic name *Kokia* is based on this native name, but the plant should not be confused with *Hibiscus kokio* (p. 48), which is found in scattered populations on all the main islands and which also carries the native name *koki'o*, presumably because it too has red flowers.

Kokia kauaiensis (Rock) O. Deg. & Duvel
Plant from Napali Kona Forest Reserve, Kauai

ʻĀHINAHINA

Argyroxiphium sandwicense

Family Asteraceae

THE *ʻāhinahina*, popularly known as the silversword, is perhaps the most famous of all native Hawaiian plants, with its spectacular appearance, enhanced by its inhospitable and lonely habitat in the "moonscape" on Haleakala, Maui, itself a magnet for tourists. The remarkable genus *Argyroxiphium*, which is found nowhere else in the world than the Hawaiian Islands, contains five species, and among them the subject of this plate is perhaps the most astonishing. The other four are less known, but *A. grayanum* (see p. 22) is closely related and bears several points of similarity. The most obvious difference between the two lies in the color of the leaves, as suggested by their Western names—silversword and greensword—the leaves having been likened to swords. The former is, in fact, a translation of the scientific name, itself derived from the Greek *argyros* ("silver") and *xiphos* ("sword").

The silvery coloration comes from the thick covering of hairs that clothe the leaves. These hairs perform two functions indispensable in the harsh conditions under which these plants grow. They help reduce loss of water through transpiration, and water is vital at windy altitudes in a cinder desert characterized by drought and quick drainage. And they reflect the fierce rays of the sun, intensified by the clear, thin air at elevations of 7,000–10,000 feet on the mountain of Haleakala where the *ʻāhinahina* grows.

This Haleakala plant is in fact subspecies *macrocephalum*. The plant originally called *Argyroxiphium sandwicense* was collected on the island of Hawaii, again at a high altitude, possibly on Mauna Loa, but today it is known with certainty only from a small area on Mauna Kea. It was first collected in 1825 by James MacRae, who had been sent by the Horticultural Society of London to obtain seeds and make collections of plants. MacRae was a passenger on HMS *Blonde*, which had been sent to carry home the bodies of King Kamehameha II and Queen Kamamalu, both of whom had died of measles while on a visit to Britain.

The silverswords of Haleakala and Mauna Kea, subspecies *macrocephalum* and subspecies *sandwicense*, respectively, are primarily distinguished by the overall shape of the inflorescence, which is much longer than broad in the latter. They may also be identified by the number of ray florets in a flower head. This difference is paralleled by the flower heads themselves, which are larger and have more florets in subspecies *macrocephalum*.

Today, the population of silverswords is limited and sporadic. Both the Haleakala and Mauna Kea subspecies are on the U.S. Endangered Species list. It appears from historic records that the silversword was found in great quantity on Haleakala only a century ago. But humans and injurious insects have combined to reduce the population: humans destroying great numbers at one time, by uprooting them before they reached the flowering stage in order to make ornaments; insects reducing the production of fertile seeds, essential for the next generation. Perhaps control of the insects could result in a good supply of viable seeds which, if dispersed, could then ensure that many more were present in the Haleakala "soil," ready to germinate whenever conditions became favorable.

Argyroxiphium sandwicense DC. subsp. *macrocephalum* (A. Gray) Meyrat
Plant growing on Haleakala, East Maui, supplemented by dried specimens

MUNROIDENDRON

Munroidendron racemosum

Family Araliaceae

MUNROIDENDRON is a genus endemic to the island of Kauai—as a wild plant it occurs nowhere else in the world. What is more, there is only a single species—although two varieties, which are no longer recognized, have been described in the botanical literature. This plant was first thought to be a *Tetraplasandra* and named as a member of that endemic Hawaiian genus, but its narrow, pendulous inflorescence, unlike anything in *Tetraplasandra*, seems sufficient on its own to give it separate status.

Munroidendron racemosum is a straight-trunked tree with light gray, smooth bark which grows to 15–25 feet. The epithet *racemosum* was given to this species because of the contrast between its apparently racemose inflorescence and the panicles, which are otherwise characteristic of *Tetraplasandra*. In fact, this dense and seemingly simple inflorescence is compound, with as many as 250 pedicels. It is probably a very condensed panicle—not a true raceme. As the fruit develops, the enlarging, globose drupes crowd together to form a dense hanging cylinder.

Although the separation of *Munroidendron* from *Tetraplasandra* is here upheld, it has been suggested that if these two genera are considered along with *Reynoldsia* (also indigenous to Hawaii and some other Pacific islands) and *Gastonia* (of East Africa, Madagascar, the Mascarenes, Malaya, Papua New Guinea, and islands as far east as the Solomons) it might be found that they should all be united in one genus. The delineation of genera in this family is difficult. Be that as it may, there is no doubt about the distinctness of this plant as a species. It is very rare, however, and listed as "Endangered" under the U.S. Endangered Species Act. Some botanists suspected that it had become extinct, but recently it has been observed in five, mostly inaccessible mesic forest localities. Fortunately, it has proved easy to propagate in cultivation, which should make it possible for interested gardeners to help safeguard this unique Hawaiian treasure.

This remarkable plant was named in honor of George Campbell Munro (1866–1963), with the suffix *dendron* ("tree") added. Noted as an early conservationist and botanical explorer, especially on Lanai, Munro was manager of Molokai Ranch early in this century and was one of the first to find this plant. He was particularly famed as an ornithologist and was the author, in 1944, of the popular and beautifully illustrated *Birds of Hawaii*, which went to a second edition in 1960.

Munroidendron racemosum (C. N. Forbes) Sherff
Plant growing at the Lawai Garden, introduced from Limahuli Valley, Kauai

ʻĀNAUNAU

Lepidium serra

Family Brassicaceae

THE genus *Lepidium* does not produce spectacular flowers, nor does it contain any ornamentals. But it has been well known from ancient times in the form of *L. sativum*, garden cress, which because of its sharp flavor has been much grown in Europe as a salad plant. Altogether, this genus contains about 150 species, distributed throughout the world, with most of them being found in the temperate regions. It is interesting, however, that on oceanic islands it has often developed endemic species, and these are often somewhat woody at the base, almost shrubby—a most unusual situation in the normally herbaceous family to which *Lepidium* belongs. On the Hawaiian chain, for example, there are two somewhat shrubby endemic species, *L. arbuscula* and *L. serra*—both proposed as candidates for listing under the U.S. Endangered Species Act—and two endemic varieties of the Pacific *L. bidentatum*. Five introduced and typically herbaceous species have also become naturalized in the Islands.

The plant from which this plate was painted was growing in a somewhat exposed position on a dry, north-facing bluff at about 1,500 feet elevation on the west flank of Waimea Canyon, Kauai. The *ʻānaunau*, as it is known to the Hawaiians, also grows in damper habitats in the area, but still at a considerable altitude and showing a preference for rocky ledges on cliffs. Under such circumstances, the leaves are somewhat thinner and the serrations much larger and more prominent. This species can reach a height of about 4 feet.

It is interesting to note that an attempt in 1973 to introduce this endemic to the National Tropical Botanical Garden and grow it there at the low altitute of the Lawai Valley failed. A subsequent attempt in a drier and apparently more appropriate rocky habitat, and without shade, was successful at first but in the end failed as well. In the wild this plant is found at altitudes of 1,500 feet or more, and it seems that even under careful cultivation it will not grow at sea level. Such difficulties emphasize the adaptation of many rare endemics to specialized habitats. The continued existence of these species is immediately threatened if these habitats are interfered with or destroyed.

The name *Lepidium* is derived from the Greek name *Lepidios*, which was used to denote garden cress by Dioscorides, a Greek physician, in his famous *De materia medica*, written around the first century of the Christian era. In Europe, this herbal was the undisputed basis of all medical information about plants until the Renaissance, when direct investigation rather than mere acceptance of "the authorities" began to regain its place as a source of knowledge. The term *lepidium* is itself derived from a diminutive of *lepis*, meaning "scale," in allusion to the small, scalelike fruit. The epithet *serra* is the Latin for "saw," in this case alluding to the sawlike edge of this species' leaves.

Lepidium serra H. Mann
Plant from the west wall of Waimea Canyon, Kauai

HĀHĀ

Cyanea coriacea

Family Campanulaceae

AMONG the entire, rich flora of the Hawaiian Islands, the family with the most remarkable representation is the Campanulaceae, the bellflower family. The subject of this plate represents *Cyanea*, one of Hawaii's five unique, endemic genera. Of these five, *Cyanea* is the largest, with no less than fifty-two species recognized in the *Manual of the Flowering Plants of Hawai'i* and ten more species (including eight previously assigned to a distinct genus, *Rollandia*) recognized since the publication of that work. Many more names have been proposed over the years, but most of these represent extreme variants within the species and have no status in terms of living populations and their biology.

Although sixty-two species of *Cyanea* are recognized, it is depressing to realize that nineteen of them are now extinct, together with a subspecies of another. Twenty unique organisms that never existed anywhere except in Hawaii are now lost forever. For some of them, all that remains are the photographs in Joseph Rock's classic 1919 monograph on the subfamily Lobelioideae. Furthermore, in the generally accepted conservation categories used by the International Union for the Conservation of Nature and Natural Resources, nine other species and one subspecies are "Endangered," one is "Vulnerable," and one "Rare." *Cyanea coriacea* is not one of these, but if the forests in the wet valleys in which it is found continue to be destroyed, then this *Cyanea*, along with the whole community of plants and animals associated with it, will be at risk.

The genus *Cyanea* is very close to *Delissea*, one of the other endemic genera in this family, and it has been suggested that the two, though distinguishable, should be treated as a single genus. But if all genera of the bellflower family are looked at as a whole, this would create other classification problems, and it is believed that the natural relationships that exist are better represented by the recognition of both *Cyanea* and *Delissea*.

The name *Cyanea* is, like so many so-called Latin botanical names, taken from the Greek, in this case *kyanos*, which means "dark blue substance." It was applied to this genus because the flowers of the first species described were said to be blue. The epithet *coriacea* describes the texture of the leaves in this species—coriaceous, or leathery. The species was first collected by the Frenchman Ezechiel Jules Rémy (1826–1893), who went on an expedition to the Pacific Islands and South America on behalf of the Paris Museum from 1851 to 1855 and collected on Kauai and Oahu in 1852. Rémy translated a little-known Island history book written by students of the Kamehameha Schools, *Ka moolelo Hawaii—Histoire de l'achipel Havaiien*, and published it in Paris in 1862.

Cyanea coriacea (A. Gray) Hillebr.
Plant collected at the National Tropical Botanical Garden's Limahuli Preserve, north shore of Kauai

KOKI'O 'ULA'ULA

Hibiscus kokio

Family Malvaceae

FOR many people, the flower perhaps most closely associated with Hawaii is the hibiscus. In this connection, it is generally the much cultivated and hybridized *Hibiscus rosa-sinensis* that comes to mind. And the official state flower is yet another hibiscus, *H. brackenridgei*. In fact, Hawaii can boast a variety of native hibiscus species, all of them endemic except *H. tiliaceus* (see p. 56). The exact number of natives varies according to the authority one follows. The *Manual of the Flowering Plants of Hawai'i* recognizes six, while early Hawaiian experts accepted a dozen or more.

With plants and animals, we are not dealing with man-made objects like nuts and bolts or airplanes, which can be sorted into distinct types. We are classifying natural organisms that are mutable, vary from place to place, and often produce intermediates through hybridization. Drawing the line between different species can be subjective. Often in the early days of floristic exploration, it happened that only the extremes from a range of one species' variation were discovered and collected. As extremes, they looked very distinct and could well have been classified as different species without any inkling, on the part of the botanist concerned, that intermediates existed between the collected specimens and therefore that the latter did not represent the whole range. This happened with *Hibiscus* (and other groups) in Hawaii. Other causes too have led to the distinctions between so-called species being set too fine, with a consequent proliferation of names. In some cases the number of species recognized in the new *Manual* may be too few, but this work provides a systematic rethinking of the whole flora—a fresh baseline from which, for example, the electron microscope and biochemical laboratory can be used in a new approach to the taxonomic problems of the Islands' flora.

The subject of this plate was drawn under the name *Hibiscus kahilii* and was collected on the slopes of Mt. Kahili above the Wahiawa Bog, the type locality for the species. But when the extent of variation in vegetative and floral characters in *Hibiscus* is considered as a whole, it is apparent that *H. kahilii* is really part of a variable and more widely distributed species for which the earliest name is *H. kokio*. It even appears to be part of subspecies *kokio*, the type subspecies. On Mt. Kahili it forms a small tree 10–15 feet tall, and its most striking feature is its bright red flowers. *Hibiscus kokio* is noted for these blossoms, and selections from the natural variability of the species have been named as cultivars, for example, 'Oahu Red' and 'Molokai Red'. It has been used for hybridization with *H. rosa-sinensis*, but the cultivated varieties that have been selected as a result of this cross seem somehow to have been treated as cultivars of the latter parent.

Hibiscus kokio, whose continued existence as a wild plant is not so endangered as is the case with several other species of *Hibiscus*, is nevertheless included as a potential candidate for listing under the U.S. Endangered Species Act.

The epithet *kokio* ("red") was taken straight from the Hawaiian name, unchanged, but it appears that a couple of variations are on record, *koki'o 'ula* and *koki'o 'ula'ula* (*'ula* meaning "bright" and *'ula'ula* "very bright") as well as *mākū*.

Hibiscus kokio Hillebr. ex Wawra subsp. *kokio*
Plant collected from Mt. Kahili, Kauai

‘ŌLULU

Brighamia insignis

Family Campanulaceae

THE Latin *insignis*, which means "outstanding" or "remarkable," is a particularly apt epithet for this plant. Its extraordinary and unexpected rosette of broad leaves at the top of a strange, tapered, thick, succulent stem marks *Brighamia insignis* out as one of the most peculiar of all the plants of Hawaii. The early Hawaiians apparently referred to this plant by a variety of names—*‘ōlulu, ālula, hāhā* or *ōhāhā, pū aupaka* or *pua ‘ala*—perhaps reflecting the interest shown in such a strange-looking plant.

The plant that is the subject of this plate had at one time been called *Brighamia citrina* (with its lemon-colored flowers), but in the *Manual of the Flowering Plants of Hawai‘i* this species is treated as part of *B. insignis*. According to the *Manual*, the genus *Brighamia* contains but two species, this one from Kauai and Niihau, and *B. rockii* from Molokai (although it is thought that it might formerly have been on Maui and Lanai as well). A few decades earlier, Harold St. John's revision of *Brighamia* suggested that there were four species, but the characters he used to separate them were of rather a minor kind—relating to the surface of the seed and the pubescence of the staminal column. Be that as it may, the status of both *B. insignis* and *B. rockii* is of concern to conservationists, and both are listed as "Endangered" under the U.S. Endangered Species Act. Many plants of both species have been raised at the National Tropical Botanical Garden from seed of plants hand-pollinated in the wild.

Brighamia plants grow naturally in open habitats on steep, often precipitous slopes where they are subject to quick drainage. Their swollen, succulent stems enable them to store and conserve water during periods of drought, such as when they are subject to strong and persistent tradewinds. This characteristic also sets the *‘ōlulu* apart from most of the Hawaiian flora, for it and the *‘ihi* (*Portulaca molokiniensis*) are the only true succulents native to Hawaii. As such, though rare in collections, they are of considerable interest to those who grow cacti and succulents as a hobby.

The particularly long and narrow tubular flowers are also unusual. This shape, combined with their white or lemon coloration, might indicate that these plants are pollinated by night-flying moths, but it would be difficult and demanding to carry out nocturnal observations in their extremely inaccessible habitats to test this theory. *Brighamia insignis* in cultivation does, however, seem to set plenty of seed when self-fertilized, so it appears not to require pollination from another plant. The long tubular flowers seem to indicate that the nearest relative to *Brighamia* might be the genus *Hippobroma* of Australia, but this is only speculation based on this one common character, one that might merely have evolved in parallel.

The genus was named after William Tufts Brigham (1841–1926), the first director of the Bernice Pauahi Bishop Museum in Honolulu and a dedicated researcher who took particular pains to collect for Asa Gray good material of the first species of *Brighamia* to be described.

Brighamia insignis A. Gray
Plant collected in northern Kauai

LOULU

Pritchardia minor

Family Arecaceae

THE coconut palm, perhaps the palm most associated with Hawaii by its thousands of visitors, is in fact a visitor itself, brought to the Islands by early Polynesians. There are, however, native Hawaiian palms as well. They are all rather special and are found only in the forests or other undeveloped, natural areas; some of them are extremely rare. To the original Hawaiians, who used strands from their leaves to weave small baskets, they were all known as *loulu*. They belong to the genus *Pritchardia*, which apart from four species from Fiji, Tonga, and the Tuamotu Archipelago is found only in Hawaii. The genus most closely related to *Pritchardia* is one found in Cuba, Guatemala, and along the Atlantic slope of mountains in Panama, called *Colpothrinax*.

Today the total number of species in *Pritchardia* is considered to be twenty-two. Although quite a number more have been described, after careful revision they cannot be maintained as distinct. It has been suggested that the Hawaiian species may well have arisen from two or more successful and chance introductions, unknown centuries or millennia ago, and then undergone gradual change. They would have been growing on an evolving and developing land; as the islands formed and became habitable, some were eventually eroded by the elements and disappeared. The offspring of the original species became separated and, in their isolation, developed their own peculiar characteristics. Today we see the results of this long differentiation, and though it may seem easy to recognize a *Pritchardia*, one needs to examine specimens closely to ascertain the species.

The plant from which this plate was drawn is one of a relatively small population in the mesic forest at about 3,500–4,000 feet altitude on Kauai. It was drawn under the name *Pritchardia eriophora* (= "wool-bearing") because of the inflorescence and flowers immersed in a dense, tan or biscuit-colored "wool," but the recent study by Robert W. Read of this genus indicates that this identification is incorrect. The plants that had been given this specific name have the densest wool of any, but as a species they are insufficiently differentiated from the earlier described *P. minor*.

Although the wool may have protected young developing inflorescences from insects and other predators that existed in Hawaii before the coming of humans, it is no defense against the horror of introduced rats. Although *P. minor* is not listed under the U.S. Endangered Species Act, tragically few seedlings or young plants are to be found in the forest today, probably because of the predation of rats or the disturbances of feral pigs. The present-day survivors need to be protected from rats, perhaps with a band of metal around their trunks, as is sometimes done with coconuts, and from pigs too—for, unless a new generation of palms can be established, extinction cannot be very far away. Four members of the genus—*P. affinis* on the Big Island, *P. munroi* on Molokai, *P. aylmer-robinsonii* on Niihau, and *P. remota* on Nihoa—are already federally listed as "Endangered." Five others—*P. napaliensis*, *P. viscosa*, and *P. hardyi* on Kauai, *P. kaalae* on Oahu, and *P. schattaueri* on the Big Island—are extremely rare and have been proposed for the same, protective classification.

The size of the fruits was much used by earlier students of *Pritchardia* to distinguish species, but it is now realized that fruit size is significant only to the extent that there are large-fruited species and small-fruited species. The latter are mostly

Pritchardia minor Becc.
Plant growing at Kokee, Kauai

restricted to low elevations, and their inflorescences are mostly lacking in hairs. There is one exception, however, which "proves the rule": *P. munroi*, known from only a single living individual in the wild, high in Molokai, and thus probably the rarest palm in the world, has small fruits and a woolly inflorescence. Fortunately, there are now some fruiting plants of this species in cultivation.

The generic name *Pritchardia* honors William Thomas Pritchard, a nineteenth-century British consul in Fiji and Samoa and author of *Polynesian Reminiscences, or Life in the South Pacific Islands* (1866). Pritchard was of great assistance to Berthold Seemann, who collected in Fiji in 1860 and wrote *Flora Vitiensis* (1865–1873).

Polynesian Colonists

HAU

Hibiscus tiliaceus

Family Malvaceae

IT will never be known with certainty whether *hau*, or *Hibiscus tiliaceus*, is truly indigenous to the Hawaiian Islands or was an early Polynesian introduction. There is no doubt that the Polynesians valued it highly for its usefulness and little doubt that it was carried by them to Hawaii along with several other important plants. Yet it might be native, already flourishing when the Polynesians arrived, for outside the Hawaiian Archipelago this hibiscus has an almost pantropical distribution and is found as a strand plant all around the globe. Its seeds have the ability to float in sea water, and months of such wetting do not appear to impair their viability. A seed shed on one of the islands thousands of miles to the south or west could no doubt have floated until, washed up by chance on a Hawaiian island, it found the right conditions for germination and seedling growth.

For the Hawaiians the main uses were derived either from the strong fibers of the inner bark or from the particularly light, tough wood. In *Plants of Hawaii National Park,* Otto Degener details several ways *hau* was employed. The bark was stripped and retted, and the long fibers dried. They were then wetted for making into binding cord, *ili hau*, and put to use in many of the same ways rope is today. The light wood, too, was used in several ways: to make outriggers and struts for canoes; to make light, tough adze handles; and to make fishnet floats. The wood is also of suitable quality to make blocks into which a hard wood, such as that of *olomea* (*Perrottetia sandwicensis*), could be rubbed ferociously to make fire.

Although natural thickets of *hau* are usually tangled, knotted, and difficult to penetrate, especially when growing in soft mud not far above the high water line, one can nevertheless find upright, straight branches as well, and these were prized to make the spears used in fireworks displays called *'ōahi*. Carefully shaped, oiled, and dried, they were set alight and hurled from precipitous cliffs. When thrown with skill, they would glide far into the air, the blazing ends reflected in the still water below.

Hau was also used symbolically. A branch set upright in the ground became an emblem in battle and was allowed to fall only when one's side was defeated. Branches were also used to signify a taboo against inshore fishing along particular beaches. Degener also mentions the use of green bark as a medicine in childbirth. *Hibiscus tiliaceus* was truly a most useful plant. It is noteworthy that the literature for native uses of tropical plants in Africa and the Americas shows that native people everywhere have used it in similar ways.

For most of us, however, it is the beauty of the flowers and the simplicity of the subtending leaves which attract. Unfortunately, individual flowers last but a day, and the strong sulfur-yellow petals (with or without a rich purplish-brown central "eye") fade to terra cotta, curl up, and drop to the ground. Furthermore, there are never sufficient flowers at any one time to make an outstanding display, but individually the blooms of *hau* are produced the year round, confirming this as a plant for all seasons.

Hibiscus tiliaceus L.
Roadside plant, Kauai

TARO

Colocasia esculenta

Family Araceae

TARO, brought to the Hawaiian Islands centuries ago by the migrating Polynesians, provided one of their main foods—for poi, the Hawaiians' staff of life, was prepared from its tubers. These tubers, which could be carried on long sea voyages, are the means by which *Colocasia esculenta* was spread by humans throughout the tropics. The species is thought to be a native of India and perhaps other parts of southeastern Asia, and from there its cultivation has extended. It reached Egypt about 2,000 years ago and also spread into the Pacific area in ancient times. It was in Hawaii, however, that the cultivation of taro, called *kalo*, reached its most sophisticated level. No fewer than sixty-seven Hawaiian varieties have been recognized. There are wet-land and dry-land taros; the former, the best known, are planted in marshy ground or in special irrigated terraces known as *lo'i*. Ancient taro fields can be seen in the Hanalei Valley, Kauai, and the remains of others are found in remote areas, now uninhabited, as for example in some of the valleys along the Napali coast of Kauai. Today, contemporary Hawaiians cultivate hundreds of acres of taro, for both domestic use and export.

All parts of the taro plant were eaten by the early Hawaiians, although the most important was the tuber—the swelling at the base of the plant. Tubers were cooked in an earthen oven, or *imu*, for two to three hours, then peeled with a stone flake or shell. In this form, taro could be used in three ways: eaten straight away; sliced and dried in the sun to produce *'ao*, in which form it could be stored and used on long voyages (and today is marketed as taro chips); or pounded, a process that led to the production of poi. Another food prepared from taro tubers started with them being raw, but cleaned and peeled. These were grated on a piece of coral, mixed with shredded coconut meat, wrapped in *lā'ī* —the leaves of ti (*Cordyline fruticosa*)—and cooked in an *imu* to produce a kind of pudding called *kūlolo*.

The young leaves and the peeled stalks of some varieties of taro were cooked as greens, like spinach, and called *lū'au* (the origin of the same word used for a Hawaiian feast, after the taro tops which were originally always served at one). The flower and its stalk, *pua kalo*, were also eaten, after the skin had been stripped from the stalk and it had been cooked. In fact, cooking is essential before any part of taro is eaten because of the crystals of calcium oxalate found in all its parts, which can cause irritation and inflammation to the mouth and throat if not broken down beforehand in cooking.

Colocasia esculenta is a member of the arum lily family. As with others in this group, what is generally called its flower is really a whole inflorescence. The cream-colored sheath, or spathe, encloses a central axis around which are borne many highly reduced, simplified and crowded individual flowers. Numerous functionally male flowers are found in an upper zone, female flowers in the bottom, and a band of sterile flowers in between. This is a common pattern in this family and is associated with mechanisms to ensure cross-pollination.

The generic name *Colocasia* is a latinization of the Greek *kolokasion*, a name used, according to the first-century Greek physician and herbalist Dioscorides, for the root of the lotus, *Nelumbo nucifera* (see p. 90), and "highjacked" for use as the scientific name for taro. *Esculenta* merely means esculent, or edible.

Colocasia esculenta (L.) Schott
Plant from a farm near Hanapepe, Kauai

KUKUI

Aleurites moluccana

Family Euphorbiaceae

ALTHOUGH the well-known *kukui*, *Aleurites moluccana*, grows like a native and appears to be part of the natural vegetation of Hawaii, it was almost certainly introduced by the Polynesians. It has, however, made itself quite at home and is today widespread on all the main islands. In the native forest, *kukui* can be recognized even at a distance by its pale green, almost silvery leaves—noticeable in particular along water courses and the bottoms of ravines. It can grow to a height of 70 feet but usually ranges between 30 and 50 feet. Its cream-colored flowers, produced in panicles terminal to the branches, are of one sex, enclosing either fifteen to twenty stamens or a two-celled ovary. The globose fruits contain either one or, less often, two hard nuts, which contain the oily kernel that made this tree so valuable to the Polynesians.

The *kukui* is the official state tree of Hawaii, so chosen because of its many traditional uses. The tree's Western name, candlenut tree, discloses its most prominent use, as a source of light. *Kukui* oil provided light in several different ways. In perhaps the best known, called *kālī kukui* (after *kālī*, "thread"), a dozen or so lightly baked nuts were threaded on the stiff midrib of a dry coconut leaflet or a sliver of bamboo. The nut at the top was lit and burned for two or three minutes. Then, when it was nearly finished, the whole candle was held upside down until the next nut was alight, at which time it was righted again while the second nut burned. When the oil from the second nut was exhausted, the process was repeated, and so on until all the nuts on the thread were used up. Torches, called *lamakū*, were prepared by stuffing bamboo with *kukui* nut kernels and other combustible materials. Probably the best light was provided by *kukui* nut oil in a stone lamp. A twist of old tapa was hung over the edge as a wick, or several wicks were used to give an even brighter light. These lamps were called *pohokano*, or *poho kukui*.

Kukui nut oil was also used in charring the center of a block of wood chosen to be made into a bowl. Oil was rubbed into the surface, over which dry combustible material was then burned. The thin layer of charcoal that remained was easier to chip out than would be the hard wood itself. The oil was also used, along with green bamboo leaves, to polish the wooden bowl at the end of its manufacture.

Because of the quality of this oil, a brief industry sprang up in Hawaii in the nineteenth century. It is reported that at one time 10,000 gallons of oil a year were exported, primarily for use in paints and varnishes. But the industry failed, presumably because the labor of collecting and transporting the nuts rendered it unprofitable.

Aside from providing oil, nuts were roasted, pounded, and mixed with sea salt to make a popular relish, *'inamona*. With their green covering removed, they were polished and made into leis (and in this form are still sold to the tourists). The soft, green outer rind produced a black dye used on tapa. Dyes were also obtained from the the bark of the roots and from the inner bark of older trees.

Various folk medicines were also obtained from this tree. The raw nut provided a purgative, and the sap was used to produce cures for skin diseases. No wonder numerous stories and myths surrounded the *kukui*—for the original Hawaiians, one of the most important of trees.

Aleurites moluccana (L.) Willd.
Tree growing by the roadside near Koloa, Kauai

SWEET POTATO

Ipomoea batatas

Family Convolvulaceae

THE sweet potato is unknown in the wild—a situation shared with a few other ancient food plants. The general opinion today is that it arose in tropical America, but investigations of the exact origin are surrounded by speculation. It is significant that the plant has a particularly high chromosome number, and the cytological makeup suggests that its chromosomes must have originally been derived from another species. *Ipomoea tiliacea* is a possible candidate as this ancestor, for it closely resembles the sweet potato and sometimes bears tuberous roots. It is native in tropical America, Africa, and Asia and, if it is the progenitor, it could have spread from Africa, west to America and east to Asia and Polynesia. The strongest hypothesis suggests that the sweet potato was derived from the native Mexican *I. trifida*, which has the same chromosome number, crosses with *I. batatas* to produce fertile hybrids, and back-crosses with either parent. *Ipomoea trifida* does not develop tubers, but their production may have been the result of a chance "sport," exploited long ago by early Aztecs or Mayans.

Whatever its specific origin, how did the sweet potato get to Polynesia? Ethnologists such as Thor Heyerdahl, famous for the voyage of the *Kon Tiki*, have used its distribution as one line of evidence to support the theory that early American voyagers crossed the Pacific. Linking the South American name *kumar* with that of *kumara*, which is widely used in Polynesia, including New Zealand, these scholars see evidence of how the sweet potato got to Polynesia. Others argue that the sweet potato could have crossed the Pacific on ocean currents as seed, carried without human aid—as suggested by J. W. Pursglove in his 1965 paper "The Spread of Tropical Crops."

However it reached the Pacific, *Ipomoea batatas* was brought to Hawaii centuries ago by the Polynesian colonists, who called it *'uala* or *'uwala*. Cuttings were planted in mounds on land too dry for taro. Depending on the variety, and there were many (over two hundred, it is said), tubers were ready for harvesting from three to seven months after having been planted. They were cooked in an earthen oven for about two hours—taking less time than taro—and were then made into either *poi 'uala*, a sour, fermented mash that was greatly relished, or *piele 'uala*, a mix of mashed tubers and shredded coconut. The tender tips of the sweet potato vine were prized as an excellent green vegetable. But only the tips were used for this purpose, at least by those with wisdom or experience, for the older leaves contain varying amounts of hydrocyanic acid.

The genus *Ipomoea*, which also contains the morning glory, is estimated to contain about five hundred species. Their characteristic habit is that of a vine. Although not every variety of sweet potato actually develops climbing shoots, most do if given the chance, and if lacking support they trail all over the surrounding ground. Many varieties have purple stems and dark green leaves, but some are paler. Their tubers also vary in color, ranging from white to orange and purple, and in the amount of starch and sugar they contain. The shape of the leaves varies greatly too, from entire and ovate or heart-shaped to five-lobed. Although their shape may vary according to the position on the shoot, particular varieties can be identified by a combination of these otherwise variable characters of the leaf, shoot, and tuber.

Ipomoea batatas (L.) Lam.
Plants growing near the roadside on Kauai

BREADFRUIT

Artocarpus altilis

Family Moraceae

THE breadfruit, with its rich green, deeply lobed, leathery leaves, is one of the most handsome and attractive of all tropical trees. Called *'ulu*, it was one of the plants transported by the Polynesians to the Hawaiian Islands in the early years of their colonization. Tradition has it that a twelfth-century chief called Kahai, the son of Moikeha, voyaged to his father's homeland, collected a plant of breadfruit from Upolu in Samoa, and returned to plant it at Kualoa on Oahu.

Somewhat different was the most famous voyage undertaken for breadfruit, that of the ill-fated HMS *Bounty* in 1789, which ended with the mutiny and the incredible voyage by Captain William Bligh and the few crew members who remained loyal to him. In an open boat with little food and few instruments, they sailed right across the western Pacific for 4,000 miles, from Tahiti to Timor. Perhaps less known is the subsequent voyage by Bligh in 1792, on which he successfully transported six hundred living breadfruit plants to the West Indies, thus achieving what had been the original aim of the voyage of the *Bounty*.

Although never as important a source of food to the Polynesians of Hawaii as taro (*Colocasia esculenta*, see p. 58), breadfruit was much eaten in Hawaii, either broiled in its skin (to protect it from the hot embers) and called *'ulu pūlehu*, or baked in an earthen oven. After it had been baked, it was mashed with the fingers, with a little water, to produce *poi 'ulu*. Alternatively, the ripe fruit was mashed and mixed with coconut cream, wrapped in the leaves of ti (*Cordyline fruticosa*), and cooked in an earthen oven to produce *piele 'ulu*. If, however, more coconut oil had been used and the breadfruit was overripe, after cooking it was allowed to cool, then cut up into slices and dried in the sun. The result was called *pepeie'e* or *paipaie'e*, and in this form the breadfruit would remain in an edible condition for a whole season, as long as it was taken out and spread in the sun again every now and then to prevent the growth of mold.

As well as being a source of food, *Artocarpus altilis* provided a light wood which, although susceptible to dry rot, was used for various products. Along with *koa* it was used to make the smaller, thinner surfboards, *alaia*, from about 3 to 9 feet in length. Short cylindrical portions of the trunk, a foot or two long, were made into hula drums. They were hollowed out from both ends, leaving a kind of concave septum about one-third of the way from one end, the basal end. Over the top, scaleless shark skin was stretched and held down by cords made from coconut fiber. The lower part was often carved in a variety of traditional patterns. Breadfruit wood was also used for many other purposes—as door posts and lintels of huts, for example, and to provide bow and stern deck planks for the larger canoes.

As with other members of the fig family, *Artocarpus* produces a white latex when cut or damaged. This latex was made into a gum that was useful in many ways. It was used in the manufacture of sticky birdlime, which was spread on baited poles to catch birds, either for food or for their feathers. The feathers were used in the manufacture of the magnificent royal capes and helmets, and the plucked and released birds, their feet washed clean with *kukui* oil, came to no harm and soon grew new plumage. Breadfruit gum was also used to caulk canoes, to seal two gourds together in the manufacture of gourd drums, and mixed with powdered pumice and clayey soil to make a kind of putty used to fill

Artocarpus altilis (Parkinson) Fosberg
Garden plant near Kalaheo, Kauai

in cracks and small holes when wooden bowls were repaired.

Botanically speaking, the male and female flowers of *Artocarpus* are borne in separate inflorescences, but in both cases the individual flowers, without petals, are greatly simplified, reduced, and crowded together. The male inflorescence is club shaped and soon withers after the pollen has been shed. The female is more or less spherical, and as the individual ovaries develop they press one against another to produce what is really a compound fruit—the breadfruit. Although the fruits of varieties grown in Hawaii, and most of Polynesia, are seedless (having been selected and in cultivation propagated clonally), there are varieties, especially in Melanesia and Micronesia, whose fruits contain seeds that may be roasted or boiled and eaten.

It is probable that New Guinea was the original and native home of *Artocarpus altilis*, but in most areas, including Polynesia and Hawaii, the seedless forms are generally of greater importance. The total number of cultivated varieties—or cultivars, as they should be termed—is not certain, but it is considerable. One of the problems in trying to sort them out is that the same cultivar may be grown under quite different names on different islands. Diane Ragone has brought together a scientific collection of more than a hundred accessions from all over the Pacific for comparative and other studies. This collection is now established at the Kahanu Garden near Hana on Maui, part of the National Tropical Botanical Garden.

The name *Artocarpus* is derived from the Greek *artos* ("bread," or "loaf of bread") and *karpos* ("fruit") and is thus a strict translation of the common name. The epithet *altilis* is Latin for nourishing or well-fed, as in a description of a bird prepared for the table.

Modern Transplants

NEW GUINEA CREEPER

Mucuna novoguineënsis

Family Fabaceae

OF the perhaps one hundred species in the pantropical and subtropical genus *Mucuna*, one of the most spectacular is the subject of this painting, the woody, climbing New Guinea creeper, *Mucuna novoguineënsis*. Vigorous in growth, it produces its vivid red, pendant racemes at least once a year and often has smaller flowerings in between as well. It is no wonder that flowering lianas such as this one have become popular ornamentals in Hawaii and throughout the tropical world.

The early popularity of members of this genus makes tracing the local origins of some cultivated species difficult, but it seems probable that one species, *Mucuna gigantea*, was distributed by the waves of the Pacific and is indigenous to Hawaii. This large liana, known as the sea bean and as *ka'e'e* in Hawaiian, is found near low-elevation streams or at the seashore on most of the major islands. Its seeds were used by early Hawaiians as a cathartic. Another species, *M. urens*, is a South American native apparently naturalized in the Islands. An indigenous Brazilian name for this latter plant is the origin of the generic name *Mucuna*.

Apart from the rather mundane name that has been applied to it, New Guinea creeper, many people will know the subject of this painting as the red jade vine, *Mucuna bennettii*, the mistaken name under which it was growing when painted by Mary Grierson. *Mucuna novoguineënsis* and *M. bennettii* both come from Papua New Guinea and are closely related species, but *M. novoguineënsis* has a truncate calyx, as can be seen in this portrait, whereas *M. bennettii* has distinct calyx teeth up to 1 centimeter long. There are also differences in the dimensions of individual flower parts, but another good way of distinguishing between the two species lies in the production by *M. novoguineënsis* of long racemes, where the flowers open progressively downward, in contrast to those in *M. bennettii*, where they all come out more or less together in a relatively short, truncate inflorescence.

When and how these plants were introduced to cultivation is not known for certain. *Mucuna bennettii* was probably the first of the genus to be cultivated. Professor R. E. Holttum has told me that in 1940 a zoologist on one of the Archbold expeditions to New Guinea gave seeds to the Singapore Botanic Gardens, of which he was then director, where they were planted and flowered a few years later. From this garden the species was distributed to Kuala Lumpur and widely throughout Malaysia—and probably from there to other countries. *Mucuna novoguineënsis*, which is a more widespread plant in Papua New Guinea, has been in cultivation at the botanical garden at Lae for many years, and it seems reasonable to suppose that it was distributed from there. Whatever the exact origin, these two striking species are without doubt great assets to tropical gardens.

Mucuna novoguineënsis Scheff.
Garden plant at Kalaheo, Kauai

DEVIL FLOWER

Tacca chantrierorum

Family Taccaceae

THE unusual *Tacca chantrierorum* is not a native of Hawaii but comes from Southeast Asia—from Assam and Bangladesh to southwest China, Vietnam, and Malaya. It was introduced to cultivation in France around the turn of the century, presumably from somewhere in Indochina, by the brothers Chantrier, nurserymen at Montefontaine in the Department of Oise.

Perhaps the most striking thing about this plant is the blackish-maroon color of the flowers, together with the similarly colored, broad outer bracts and ten or twelve drooping, filiform inner bracts. This color has given rise to one common name, devil flower; the broad, dark-colored bracts have inspired another, bat flower. As is usual with flowers of this color, there is an associated smell of rotting flesh which attracts flies. The insects crawl inside the flower but find it difficult to emerge, because the helmet-shaped stamens and the lobes of the stigma are in the way. But when their efforts to escape are successful, and if they then visit another flower, cross-pollination may be the happy outcome.

The genus *Tacca* contains ten species and has a pantropical distribution. In Hawaii, as well as *T. chantrierorum*, which is known only as a cultivated plant, there is a naturalized species with divided leaves, *T. leontopetaloides*. The early Polynesians, who introduced this plant to Hawaii, called it *pia*; it is also known as Polynesian arrowroot. Presumably these ancient voyagers brought it with them because the tuber is a good source of starch. The tubers are, however, very bitter, and so they had to be prepared by grating, soaking, and repeated washing in fresh water—efforts made only in times of famine when other food alternatives were limited. The resultant starch was mixed with coconut oil and baked or boiled to produce a pudding, *haupia*. It is said that any starch not immediately needed was shaped into small cakes, dried in the sun, and stored.

The botanical name *Tacca* is based on *taka*, the native Moluccan name for another species of this genus. Although the epithet appears as *chantrieri* in many reference books and was in fact spelled that way by Edouard-Françoise André himself when he first described this species, because he attributed it to "MM. Chantrier fréres," in accordance with the rules of nomenclature it must not remain in the singular, *chantrieri*, but unfortunately must be changed to the less euphonious plural, *chantrierorum*.

The moth depicted in this plate is associated with *Tacca* only coincidentally—through the shared color. It is the black witch moth, *Ascalapha odorata*, a native of the West Indies which was introduced to Hawaii early in the century. With a wing span of over four inches, it is the largest lepidopteran to be met with in the Islands. Its caterpillars feed at night on the monkey pod tree, *Samanea saman*, and on some other leguminous trees.

Tacca chantrierorum André
Plant in the Lyon Arboretum, Honolulu, Oahu

ANGEL'S TRUMPET

Brugmansia × *candida*

Family Solanaceae

THIS spectacular member of the nightshade family is often grown under the names *Datura candida* or *D. arborea*, but it is now generally recognized that *Brugmansia*, which for a long time was treated as a section of *Datura*, warrants recognition as a distinct genus on its own. There are several technical points that have to be considered when deciding whether *Brugmansia* and *Datura* should be treated as separate genera, but put simply, species of *Datura* (which contains the thornapple, or jimson weed) are annuals and have flowers that are held upright, whereas species of *Brugmansia* are woody perennials with flowers that hang down. This difference is clearly captured in the Hawaiian name for this plant, *nānāhonua*, or "gazing earthward." The flowers, about 10 inches long with five long-pointed lobes, are also the obvious referent of the common name in English-speaking countries, the evocative angel's trumpet.

Plants that have arisen by crosses between three species of *Brugmansia* are known in cultivation, although the hybrid featured here is the most familiar. Furthermore, within *B.* × *candida* three or four named varieties, or cultivars, have been selected. The subject of this painting is most likely that called 'Grand Marnier'.

All members of *Brugmansia*, about seventeen species, are native to South America, but there has been uncertainty about the original home on that continent of the subject of this plate. It seemed not to be known in the wild. But now it is realized that it is of hybrid origin, although anciently so, and represents a cross between *B. aurea* and *B. versicolor*. This realization explains why fruit is rarely set, although the plant is widely distributed and grown in South and Central American gardens, where it goes under the name floripondio. According to Otto Degener, this ornamental reached Hawaii in 1825 on a ship from Rio de Janiero.

The angel's trumpet has an attractive, fragrant and musky scent, especially in the evening. The leaves, and probably all parts, of this plant are highly poisonous. Many members of the Solanaceae family are noted for their alkaloid poisons: nightshade, belladonna, tobacco, and mandrake, for example. This quality explains why species of *Brugmansia* are prized by certain indigenous South Americans for their hallucinogenic properties and used in their religious ceremonies.

The genus was named in honor of Sebald Justinus Brugmans (1763–1819), a Dutch botanist and Professor of Natural History at Leiden. The epithet *candida* means "shining" or "pure white," a description that fits those forms of this hybrid which have white flowers.

Brugmansia × candida Pers.
Plant in the Allerton Garden, Lawai, Kauai

BIRD OF PARADISE

Strelitzia reginae

Family Strelitziaceae

IN South Africa, where *Strelitzia reginae* is native, it is known as the crane flower; in cultivation throughout the world, the name more generally used is the bird of paradise. Both names allude to elegant birds, but the latter name seems more appropriate for this striking subject, and more in keeping with its stiff flamboyance. The bird of paradise can be grown in climates like that of Hawaii with relative ease, provided it has a reasonable supply of water. This, coupled with its exotic image, ensures its popularity. Its value as a cut flower adds to this success and, on an increasing scale, it is even grown commercially.

Behind the decorative appearance of the plant lies an intriguing structural adaptation for cross-pollination. In the wild, the bird of paradise is pollinated by African sugar-birds and sun-birds attracted by nectar and the three orange sepals with contrasting blue, arrow-shaped lateral petals (the third petal is much reduced to a blue flap at the base and protects the nectar). The bird alights on the perch formed by the sheaths of the two petals and in doing this depresses them so that stamens, which were hidden and held between them, are forced out and rub the bird's breast with pollen. Later, after the stamens have withered, the stigma becomes receptive and protrudes, ready to brush against other birds visiting the flower, and in its turn receives a supply of bird-carried pollen, thus effecting cross-pollination. As a further aid, the pollen grains of *Strelitzia* are not dry and dusty as in the majority of plants but are somewhat sticky, increasing the probability that they will adhere to birds' feathers.

Birds are also important for seed dispersal of *Strelitzia*. The shiny black seeds have, at their base, a fluffy tuft of orange hairs that contain nutrient fats. These are sought after and eaten by the birds, who leave the seeds to fall wherever they may.

The origin of the botanical name for this plant is not without interest. *Strelitzia reginae* was introduced from the Cape of Good Hope to the Royal Gardens at Kew in 1773 and caused a minor sensation when it eventually flowered. Sir Joseph Banks, who was unofficial director of Kew under King George III, named it after his queen—hence the epithet *reginae*. Queen Charlotte, being keenly interested in gardens, did much to develop the Royal Gardens at Kew. Nor was the specific name honor enough; Banks also linked this spectacular subject to her through the generic name—for she had been born Princess Charlotte Sophia of Mecklenburg-Strelitz.

The butterfly flitting about the bird of paradise is a monarch, *Danaus plexippus*, a species introduced to Hawaii and now well established there.

Strelitzia reginae Banks ex Aiton
Garden plant near Kalaheo, Kauai

NIGHT-BLOOMING CEREUS

Hylocereus undatus

Family Cactaceae

NO cactus is native to Hawaii, but this plant has been growing in the Islands for over a century. Otto Degener's 1932 *Flora Hawaiiensis* account of this plant and its introduction can scarcely be bettered: "Around 1830 the night-blooming cereus and a number of other plants were put on the brig *Ivanhoe* while that vessel was stopping at a Mexican port on the way from Boston to Canton. Most of these plants died before the vessel reached the Hawaiian Islands. As they were being thrown away Captain Charles Brewer, then first officer, rescued the night-blooming cereus because it still showed signs of life. This specimen was planted in Honolulu and flourished. Cuttings from the original importation were early planted in Wailuku, Island of Maui, and elsewhere. Now the plant is a common ornamental grown over rocks and walls in the drier districts at lower elevations; occasionally escaping. The cereus-covered stone wall around Punahou School, Honolulu, is probably the most famous hedge in the Islands. . . . The plants bloom between June and November, usually producing sporadically, according to the weather, 4 or 5 magnificent floral displays interspersed by smaller ones. Because of the beauty of the flower a law has been passed prohibiting their picking until after 10 o'clock at night, thus giving everyone opportunity to see the plants at the height of their glory—a sight that should not be missed. By placing the opening flower in a refrigerator early in the evening, its blooming can be arrested so that the flower will remain open part of the following day even when removed from the cold." —The famous hedge is now more than 150 years old. One wonders if the 10 o'clock prohibition is still in force.

Originally called *Cereus undatus*, which persists in the common name for this plant, the night-blooming cereus was transferred to the genus *Hylocereus* by N. L. Britton and J. N. Rose in their classic work *The Cactaceae*, published in four volumes between 1919 and 1923. The epithet *undatus* means wavy, in allusion to the scalloped margins to the wings on the stems. The generic name comes from attaching the Greek *hyle* ("a wood") to the name "cereus," so meaning "forest cereus." The Hawaiians called it *pānini-o-Ka-puna-hou* ("Punahou cactus"), a reference to the famous hedge at Punahou School mentioned by Degener.

This species is native to Mexico and is planted as an ornamental in most tropical and subtropical countries with a dry enough climate—and it has escaped in many of them. Because it is self-incompatible and needs to be pollinated by another genetically individual plant, it is usually propagated vegetatively—that is, from pieces of a single plant. Thus, most so-called populations are part of the same clone, and seed is rarely set. Still, in its native Mexico it is said that its oblong, red or white juicy fruits are refreshing to eat and often to be found in the native markets of Yucatan.

Hylocereus undatus (Haw.) Britton & Rose
Plant growing in the Allerton Garden, Lawai, Kauai

QUEEN EMMA LILY

Crinum pedunculatum

Family Amaryllidaceae

THIS well-loved ornamental, almost certainly the true Queen Emma lily, is said to have been one of the favorite flowers of that much-loved queen, the consort of King Kamehameha IV. Deeply religious and a strong Anglophile, Queen Emma was closely involved in numerous charitable works. To take one example with an unexpected botanical link, she had much to do with the foundation of Queen's Hospital in Honolulu in 1860, a facility for sick and destitute Hawaiians. The botanical link comes with the first physician at that hospital, Dr. William Hillebrand, who was also the author of the classic *Flora of the Hawaiian Islands*, published in 1888 and only recently superseded as the most comprehensive study of Island flora.

Queen Emma was also noted for her keen interest in flowers and gardening. As an often melancholy widow (having adopted the name Kaleleonalani, meaning "the flight of the chiefs," after the early deaths of both her son and her husband), she lived just above Lawai-Kai in 1870 and 1871. The remains of her wooden cottage are now to be found in the mouth of the valley. Almost a century ago, the cottage had been lowered from its original cliff site onto foundations next to the Lawai Stream, where it faced out over the sea across the wide palm-shaded strand. In retrospect the move was unfortunate, because in 1992 the storm surf of Hurricane Iniki overwhelmed that beach, and the cottage.

There has been considerable speculation over which plant was the real Queen Emma lily. In the first edition of *In Gardens of Hawaii* (1948), Marie C. Neal identified the lily as *Crinum amabile*. But by the time of her revised edition (1965) she had changed her mind and called it C. *augustum*, although using the same line drawing to illustrate it. Important evidence has been provided by Mr. Hideo Teshima, who grew up in the Lawai Valley and worked in the Allerton Garden there most of his life. According to him, the lily now growing by the Diana Fountain, depicted in the plate, was transplanted there from its original location by the "Jungle Garden" near the beach. He also believes that C. *pedunculatum*, which is abundant at Lawai-Kai, was the species grown there in Queen Emma's time and the species she favored.

The taxonomy of the genus *Crinum* is in a state of disarray. There is need for a wide-ranging and complete revision, using living plants from throughout its geographic range. This is not the place to express firm opinions on the different species that have been recognized, but it seems possible, because they are sterile, that plants known as C. *amabile* and C. *augustum* are of hybrid origin, and that neither were growing at Lawai-Kai in Queen Emma's day. Apparently C. *pedunculatum* was there, and it has often been overlooked that this species, which generally has pure white flowers, also has a variant in which the floral tube is dark purple and the outside of the "petals" are flushed with purple. This is the plant painted by Mary Grierson.

Crinum pedunculatum R. Br.
Plant growing in the Allerton Garden, Lawai, Kauai

STARFRUIT

Averrhoa carambola

Family Oxalidaceae

THE name "starfruit" may be unexpected, but cut the ridged fruit of this plant across in slices and it immediately becomes appropriate. In fact, ripe fruit sliced in this way is often used in salads or as a garnish. In addition, the fruit is also stewed and used in tarts or curries. It is, however, somewhat acidic, so in some countries it is cooked with sugar or honey, or made into jam, chutney, or pickles. The juice of the fruits, provided they are ripe, makes a delightfully refreshing and cooling drink, rich in vitamin C, which is easily made with a blender. Another drink made from the starfruit has been recommended as an antidote to hangovers. Remarkable though it may seem, it was noted for this therapeutic property centuries ago, by Georg Eberhard Rumphius (1627–1702), in his classic *Herbarium amboinense*, one of the earliest books on tropical plants.

Starfruit, also known as carambola, is generally thought to be native to Sri Lanka or the Moluccas and has been widely cultivated in tropical Asia for centuries. Today it is grown worldwide in the tropics. The flowers are small, red or pink, and may themselves be used in salads or made into preserves. But it is for the fruit, particularly characterized by the five or six longitudinal ridges that run its full length, that it is generally known and grown. The skin is waxy, in shades of orange or yellow, and when ripe the flesh is yellow.

Although there are several named varieties, they fall into two groups—one with smaller, very sour but richly flavored fruit, the other with larger and sweeter fruit but only a mild flavor. To a great degree the difference in taste depends on the amount of oxalic acid they contain, and for this reason starfruit is not recommended to those who need to avoid acidic food. In fact, because of the presence of so much oxalic acid in the plant, its juice can be used to remove metallic corrosion. This has led to its reputation for cleaning and polishing metals, especially brass. The juice can also be used to remove rust stains from white linen. Indeed, the whole family of plants to which *Averrhoa* belongs, the Oxalidaceae, is noted for the acidity of its members. Oxalic acid is itself named after plants of the genus *Oxalis*, the large, type genus of the Oxalidaceae. The family is overwhelmingly herbaceous, and the two species that constitute the genus *Averrhoa* are the only arborescent members. They can form small trees up to 30 feet tall.

The name *Averrhoa*, which dates as a botanical name from 1753, commemorates the Moorish physician and philosopher ibn-Rushd (ca. 1126–1198) under the name by which he was known in the universities of Europe—Averroës. Ibn-Rushd's many commentaries on Aristotle's writings became an integral part of contemporary Latin philosophy and science, with influence lasting well into the seventeenth century. The specific epithet *carambola* dates even earlier. It was first recorded by a Portuguese physician, Garcia da Orta, who lived in Goa, the earliest European colony on the west coast of India, from 1534 to his death in 1570. Garcia da Orta used the name "carambola" in 1563, in his *Coloquios dos simples e drogas he cousas mediçinais da India*, the first book on Indian medicinal drugs and plants. He wrote, "Carambola, for so we call them in Madras, and we stick to the Malayalam names." This was not, however, Malayalam but Portuguese, and at times the name has been applied to the fruit of more than one plant.

Averrhoa carambola L.
Plant collected at the National Tropical Botanical Garden, Lawai, Kauai

SHELL GINGER

Alpinia zerumbet

Family Zingiberaceae

SHELL ginger, *Alpinia zerumbet*, has been in cultivation in the West for about two hundred years. It is probably native to one or more of the islands of Southeast Asia, but it is widely popular today in most tropical gardens. After escaping from cultivation, several other members of the ginger family have become naturalized in Hawaii. At least as yet, however, shell ginger has been limited in its invasiveness. And thanks to its attractive flowers, it is still often found as a garden plant.

As well as being attractive, shell ginger flowers are unusual botanically. The flowers of most plants are symmetrical. Some, like the rose, are radially symmetrical—they can be cut into two equal halves down through the center on any radius. Others, like a pea flower, are bilaterally symmetrical—only two equal halves can be obtained by a vertical cut. In the genus *Alpinia* and the whole ginger family, in contrast, the flowers are asymmetrical and the flower parts highly modified. This can be seen by examining the flower closely. The outer, white portions tipped with pink are bracteoles and perianth segments ("petals"), and the yellow shell-like lip, red within, actually represents two stamens, here sterile and petaloid, which have taken over much of the function of attracting pollinators. The pollen is produced inside the flower by the single fertile stamen.

None of the other members of this family in Hawaii have suffered such a complex mix-up over their scientific names. This situation has now been sorted out by two authorities on this family, B. L. Burtt and R. M. Smith. Although the correct name is *Alpinia zerumbet*, this attractive plant may often be found referred to in the literature as *Alpinia nutans*, *A. speciosa*, *Catimbium speciosum*, or *Zerumbet speciosum*. It should not, however, be confused with *Zingiber zerumbet*, a wild ginger generally thought to have been an early Polynesian introduction to Hawaii. Although shell ginger is perhaps its best-known common name, it is sometimes referred to as the shell flower, and the Hawaiian name is *'awapuhi luheluhe*, which may be translated as "drooping ginger."

The meaning and origin of the scientific name may also be of interest. The generic name commemorates Prospero Alpino (1553–1616), Professor of Botany at Padua in Italy, the scene of one of the first, if not the very first, botanical gardens. Alpino lived in Egypt for three years, as physician to the Venetian consul, and published the first books on Egyptian plants and medicine. He was also the first to deduce that there are different sexes in plants by observing that the female date palm does not bear fruit without "dust" (pollen) from a male palm first being sprinkled on its flowers. The specific epithet *zerumbet* is of Arabic origin and appears to have been first used by Serapion, an eleventh-century physician and scholar, for the aromatic roots of a member of this family.

Alpinia zerumbet (Pers.) B. L. Burtt & R. M. Sm.
Plant growing in the Allerton Garden, Lawai, Kauai

CALICO FLOWER

Aristolochia littoralis

Family Aristolochiaceae

COMMON names come from various sources. Who first called the subject of this painting the calico flower is not recorded, but one may assume that the name alludes to the flower's color and pattern, for they resemble those of certain types of printed cotton cloth know as calico, which itself is named after the ancient city in India whence it first came, the city of Calicut.

The medieval "Doctrine of Signatures" lies at the back of the name *Aristolochia.* This doctrine held that a medicinal herb showed from its shape, coloring, and other characteristics the disease or ailment for which it was the remedy. In the case of *Aristolochia,* the curved form of the flowers of the European *Aristolochia clematitis* suggested a human fetus in the correct position for birth, and so it was believed that as a medicine this plant eased parturition. The name followed naturally, from the Greek *aristos* ("best") and *lochia* ("childbirth"). In fact, the folkname for this species is birthwort, and curiously enough it is now known that it contains an alkaloid that can act as an abortifacient.

The epithet *littoralis* refers to a shoreline environment, in this case the banks of the Parana River in Paraguay, from which it was first described. *Aristolochia littoralis* is native to the general area where Argentina, Brazil, and Paraguay meet, and it has sometimes escaped from gardens and become naturalized here and there in Florida. It is also reported to be naturalized on Oahu, though sparsely. In cultivation, the calico flower is frequently but incorrectly referred to as A. *elegans*.

Aristolochia is a mainly tropical genus and contains about three hundred species, most of which are climbers. Many, like the calico flower, bear spectacular but curious flowers and, as a consequence, are often cultivated by those who like to grow unusual plants. The fruit is unusual too. When mature it hangs down, splits open from the top, and scatters the flattened, winged seeds like a censer.

The flowers are predominantly dark purple or reddish brown, with spotting or intricate patterning. As is typical of flowers of this coloration, they usually have a strong fetid smell that attracts flies and other insects. Such flowers in this and other species of *Aristolochia* act as flytraps. Insects, particularly small flies in the case of A. *littoralis*, are attracted and land, only to find themselves in a slippery zone with downward-pointing, hinged hairs. The flies are quite unable to prevent themselves from sliding down into the bulbous part of the flower, where the reproductive organs lie. There they appear to be trapped, but they eventually escape when the hairs wither. This takes place, however, only after the stamens have matured and shed their pollen over the flies, usually on the second day after the flowers have opened. Thus these same insects, by repeating their folly, effect cross-pollination.

Aristolochia littoralis Parodi
Plant growing at the National Tropical Botanical Garden, Lawai, Kauai

JACK FRUIT

Artocarpus heterophyllus

Family Moraceae

THE most striking thing about the jack fruit, *Artocarpus heterophyllus*, is the size of its fruit—from 1 to 3 feet long and 10 to 20 inches in diameter, weighing at the extreme as much as 70 pounds. The fruits hang down from the larger branches and even from the very trunk of the tree itself. As in the case of the related breadfruit (see p. 64), the developing individual ovaries of *A. heterophyllus* press one against another and fuse, producing what is really a compound fruit. Another familiar example of this phenomenon is the pineapple.

Widely cultivated in tropical Asia, the jack fruit's original native country cannot be stated with certainty. It seems very probable, however, that it grows wild as a native in the rain forests of the Western Ghats, a mountain range in peninsular India. The name "jack," or "jak," comes from *chakka*, the name for this plant in Malayalam, the language of the state of Kerala, which encompasses the southern Western Ghats. There is no doubt that it has been cultivated for centuries, for it was even mentioned by Pliny the Elder (A.D. 23–79) in his *Historia naturalis*, and it was probably the tree "which is very large and has wonderfully sweet and large fruit" referred to by Theophrastus even earlier, about 300 B.C., in his *Enquiry into Plants*. Since the Renaissance, however, when the first *Artocarpus* was described in Garcia da Orta's *Coloquios dos simples e drogas he cousas mediçinais da India* (1563) or Caspar Bauhin's *Pinax theatri botanici* (1623), there has been much confusion over the name of the jack fruit, the breadfruit, and a third species, the chempedak.

Today, one can find two main types of fruit—a smaller one with soft and very sweet flesh, and a larger one with a crisper, firmer texture. But there are many different jack fruit varieties in India, Ceylon, and Indochina. The soft flesh may be eaten raw, boiled, or fried. The seeds, however, are poisonous until cooked, usually by roasting, after which they are used in curries.

As well as for its fruit, *Artocarpus heterophyllus* is prized for its timber. In particular, the heartwood of trees fifty or more years old provides durable yet easily worked wood that can be used in the manufacture of furniture and musical instruments, for example, as well as in building construction. At first the wood is a pale lemon color, but with exposure to the air it darkens with time to resemble mahogany. It is reported that the wood, when boiled, produces a yellow dye used to color the robes of Buddhist monks in Burma and Thailand.

Some readers may wonder at the epithet *heterophyllus*, which means "diversely leaved" (from *heterus* and *phyllon*), for a glance at the plate will show that this plant has entire, more or less uniform leaves. In fact, the name is a reminder of the leaves on juvenile plants and young shoots and suckers, which are narrow and have two or three lobes in contrast to the broader, unlobed adult leaves.

Artocarpus heterophyllus Lam.
Plant growing at Grove Farm, Kauai

WOOD ROSE

Merremia tuberosa

Family Convolvulaceae

BECAUSE of its attractive flowers and remarkable fruits, the wood rose, *Merremia tuberosa*, has been in cultivation for a long time in many parts of the tropics. In addition to these attributes, this native of South America is a rampant climber, and so its use as a screen has added to its popularity. Unfortunately, the attractive, funnel-shaped, scented yellow flowers, which open early in the morning, do not last the day. In Hawaii as in other places, the wood rose has escaped and now, without the help of humans, it flourishes and reproduces itself here and there, particularly in disturbed sites, often near the roadside, on Kauai, Oahu, Maui, and Hawaii.

The name "wood rose" can be explained by the appearance of the dried mature fruits, which look like flowers skillfully and delicately carved in wood thanks to the five enlarged, petal-like sepals. These fruits are frequently used in dry bouquets and are often incorporated in ornamental arrangements bought by tourists as mementos of their holiday in the tropics.

Merremia contains about eighty species distributed throughout the tropics of both hemispheres. It is very close to genus *Ipomoea*, also of the morning glory family. It seems that, as they are at present classified, the only consistent difference between the two genera is that species of *Merremia* have smooth pollen grains whereas those of *Ipomoea* are spiny. This character can be observed with a strong magnifying glass, so identification as to one or the other genus is easy, but this seems a slender distinction for separating two genera. Still, no one has yet found the time or opportunity to undertake the formidable task of reviewing the five or six hundred species involved and attempting to produce a sounder, more scientific classification. One may, however, note that all the species of *Merremia* appear to have yellow or, less commonly, cream or white flowers, whereas only four species of *Ipomoea* are known to be yellow flowered.

One other species of *Merremia* is also familiar in the Hawaiian islands and might be considered naturalized: M. *aegyptia*, the hairy merremia, is also found in dry, disturbed sites but can be distinguished from M. *tuberosa* by its palmately compound leaves and reddish hairs.

The genus name *Merremia* commemorates Blasius Merrem (1761–1824), Professor of Natural Science at the University of Marburg. The Hawaiian name for this plant is *pilikai*. The specific epithet *tuberosa* has been given to it because of the tuberous roots this plant develops. In times when jalap, a purgative produced from the ground tuber of *Ipomoea purga*, was sold in pharmacies, ground M. *tuberosa* root was sometimes used as an adulterant.

Merremia tuberosa (L.) Rendle
Plant growing by the roadside at Koloa, Kauai

SACRED LOTUS

Nelumbo nucifera

Family Nelumbonaceae

THERE are two species in the genus *Nelumbo*: one, the subject of this painting, is native to the warm and tropical parts of Asia, from India to Japan and northeast Australia; the other is indigenous to North America, from southern Ontario to Minnesota, south to Florida and Texas. Both are aquatic plants, growing in mud under shallow, quiet or slow-moving water. *Nelumbo nucifera* has rose-red to white flowers; in the American species, *N. lutea*, the flowers are yellow. In Hawaii the two species have been hybridized.

The name "sacred lotus" reflects the status of this flower in India, Tibet, China, and Japan, particularly in its association with Hinduism and Buddhism. Much mentioned in early Sanskrit writing, the lotus is woven into mythological fables and religious ceremonies. For centuries it has been the symbol of purity and truth. Buddhist saints are often depicted as seated on a lotus flower, which, with its wheel-like form, is symbolic of reincarnation and the perpetual cycles of existence. The lotus is also widely held to be a symbol of fertility, and in this connection it is interesting to note that the seeds are remarkable for their extreme longevity: an apparently authenticated case has established a viability for them of about 3,000 years.

It may cause some confusion that the name "lotus" has been used for a wide variety of plants. The ancient Greek *lotos* was applied to several fodder plants related to clover, and the modern botanical genus *Lotus* contains about one hundred such species. The lotus in the title of Tennyson's poem "The Lotus-Eaters" is *Zizyphus lotus*, a small tree of southern Europe and North Africa which produces olivelike but sweet fruits. From these a wine and a kind of bread were made, and according to Homer many of Odysseus' sailors lost all desire to return home after eating the fruit.

The fruit of the sacred lotus is also edible, starchy and nutritious, although the extremely bitter-tasting embryo must first be removed. The underwater stems, or rhizomes, which also contain starch, are eaten as a vegetable in India, Indochina, and China. By preference they are eaten when young, for when old they act as a diuretic. In fact, in Indian and Chinese medicine every part of this plant is used for many different treatments. Another interesting product comes from the spiral vessels in the leaf stalks, which when extracted and dried are apparently used as wicks for sacred lamps in Buddhist and Hindu temples.

Aside from sacred lotus, this plant is also known as the Indian lotus and the lotus bean. The botanical name *Nelumbo* has been taken almost unchanged from the Sinhalese word *nelum*, or *nelumbu*, and the epithet *nucifera* comes from the Latin *nux* ("nut") and *fero* ("I bear").

Nelumbo nucifera Gaertn.
Plant growing at Allerton Garden, Lawai, Kauai

BANANA POKA

Passiflora mollissima

Family Passifloraceae

ONLY too frequently on oceanic islands, and Hawaii is no exception, attractive plants that have been introduced to gardens escape and run riot in the native forest. The indigenous flora of such islands is susceptible to competition, for in its pristine state the vegetation is limited to the relatively few plants that have managed to reach the islands, colonize, spread, and diversify. A factor that adds to this susceptibility on Hawaii is the lack of natural predators—and certainly of native herbivores—so that when plants and animals already adapted to mainland competitive conditions are introduced, they easily dominate and destroy the indigenous species. Banana poka is one of these "vegetable terrorists."

Banana poka was introduced to cultivation in Europe, under glass in 1843, because of its attractive flowers, and it was soon taken to gardens in the tropics. In Hawaii, according to the Degeners, it was introduced to hide an outhouse in the early twentieth century. It was first collected on Hawaii in 1926 by L. H. MacDaniels at 3,000 feet altitude, and by 1970 it had become a serious pest. On the island of Kauai too, it had arrived in the forests around Kokee, again at an altitude of about 3,000 feet. The vine reaches up to the tops of the highest trees, cutting out their light, and hangs down in curtains, overgrowing and smothering the trees and shrubs beneath it. The beautiful flowers are followed by elongated ellipsoid fruits that turn yellow when ripe, giving the plant its common name on the Islands. These are much sought after by alien fruit-eating birds and feral pigs, both of which spread the seed. The very size of the infestations and ruggedness of the terrain on which banana poka grows rule out any possibility of eradicating the plant by weeding; by 1981, over 200 square miles of native Hawaiian forest had been invaded. Attempts are being made at biological control; a Colombian caterpillar that feeds on the species in its native home is being tested, and other natural predators are being sought in that country too.

Banana poka has usually been identified as *Passiflora mollissima*, a native of Venezuela and Colombia, south to Peru and Bolivia. It is under this name that it generally appears in a botanical context, although elsewhere it is identified as *P. mixta*, a closely related species with a similar native distribution. Careful examination shows, however, that it is neither of these species. It is closest to *P. mollissima*, and it now seems probable that it arose in the wild by introgressive hybridization between this species and some other. Strictly speaking, it should not be called *P. mollissima* if it is a hybrid, and since it was almost certainly introduced from a cultivated plant it is appropriate to treat is as a cultivar; as such it has formally been given the cultivar name 'Banana Poka'.

Not only in Hawaii has this plant become naturalized. It is also a pest in New Zealand and East Africa. In Kenya there are two variants, one that seems to coincide with the true *Passiflora mollissima* and another that coincides with banana poka and, again, is suggested to be of hybrid origin. Certainly the possibility of a hybrid origin (and numerous interspecific crosses have been made in *Passiflora*) would help to account for its vigor. More work must be done on the taxonomy of this plant to be positive of its identification—work that could be basic to attempts to introduce biological controls.

Passiflora mollissima (Kunth) L. H. Bailey 'Banana Poka'
Plant by the roadside at Kokee, Kauai

LILIKO'I

Passiflora edulis

Family Passifloraceae

ALTHOUGH not as serious a pest as its relative the banana poka (p. 92), *Passiflora edulis* is nevertheless spreading gradually in wild habitats—aided by non-native birds that eat the fruit and disperse the seeds. This species is said to have been first introduced to Hawaii at Lilikoi Gulch on East Maui, where it escaped. Not possessing a native name, and being a completely new plant to the Islands, it came to be called *liliko'i* after the locality, a name that has stuck with it ever since. By the time William Hillebrand's *Flora of the Hawaiian Islands* had been published, in 1888, it was reported that the *liliko'i* had already escaped into the woods of East Maui, and not long after, by 1897, Amos Arthur Heller recorded it as naturalized on Kauai.

In the normal form of this species, the egglike fruit is purple or greenish-purple when ripe (see detail, right), but there are plants of a different form in which the ripe fruit is yellow or yellow tinged with red, as depicted in our painting. These fruits contain a yellowish-orange edible pulp that is tart but somewhat sweet and fragrant, and it is for this that the plant has been cultivated, sometimes commercially. The pulp is actually an aril surrounding each seed. Along with the juice, this pulp is good in fruit salads and makes a refreshing drink and a flavorsome dessert. Forma *flavicarpa* has larger and distinctly better tasting fruit, and it was introduced by the Hawaiian Agricultural Station in 1923 for this reason. All would be fine if the plants could have been confined to cultivation, but they have escaped and added yet one more alien to the heavily invaded Hawaiian forests.

A familiar common name for this plant in English-speaking countries is the passion flower, and its fruit is well know as passion fruit. Several other members of the genus are cultivated widely in the tropics for their fruit: examples include *P. quadrangularis*, the giant granadilla; *P. ligularis*, the sweet granadilla; and *P. maliformis*, the sweet calabash.

The name *Passiflora* ("passion flower") alludes to the magical association of the parts of the flower to Christ's Passion. *Passiflora edulis* is a native of Brazil, and when the Spanish first encountered another species with similar flowers (there are over four hundred species in the genus *Passiflora*) during the time they were conquering that subcontinent, their missionaries saw it as a sign and miraculous aid in the conversion of the natives to Christianity. It was claimed that the five sepals and five petals represented the ten apostles present at the cross, the corona (in some species appearing as though in two whorls) the crown of thorns and the halo, the five stamens the five wounds, the three stigmas the three nails, and the tendrils and lobed leaves the whips and hands of the persecutors.

Passiflora edulis Sims f. *flavicarpa* O. Deg.
Plant on the roadside at Lawai, Kauai

KOSTER'S CURSE AND GLORYBUSH

Clidemia hirta and *Tibouchina urvilleana*

Family Melastomataceae

THE subjects of this painting have two things in common: they both belong to the melastoma family, and as far as Hawaii is concerned they are both noxious, introduced weeds.

According to a note from Otto Degener, along with a specimen he had collected in 1962, *Clidemia hirta* was "several decades ago purposely planted out from Wahiawa nursery." Certainly by 1941 it was reported from the Koolau Range on Oahu, although the infestation was not then very extensive. But the plant spread rapidly, and its effects on native vegetation have been devastating. In 1954, *Liothrips urichi*, a small sucking insect that attacks the growing shoots of this plant, was released in an attempt to stop the invasion. This tactic had proved to be an effective biological control in Fiji, where *C. hirta* was already a pest. But it turned out that the insect preferred open habitats. It prevented *Clidemia* from invading pastures and croplands in Hawaii, as it had done on Fiji, but it evidently shunned shady conditions, and so the alien plant continued to spread in forested areas. And it spread vigorously, so much so that by the early 1960s it could be found throughout most of the Koolau Range and has been described there as a "ubiquitous forest weed." In the early 1970s, populations of the pest were also reported from the islands of Hawaii and Molokai. It is now also known from at least a dozen populations on Kauai which are spreading rapidly, especially after Hurricane Iniki in 1992. Other biological controls are being sought, for it is now too late on these islands to eradicate it mechanically or by the destruction of individual plants. But a close lookout for it should be kept on the other islands in the archipelago, for these simpler methods, even though they may be labor intensive, are the most effective means of eradicating it when the plants are young and before there is any chance of their setting seed.

Clidemia hirta is mainly spread by birds who eat the small, blue-black, juicy berries that develop from the rather insignificant white flowers. It is ironic that the bird that is perhaps responsible for most of the seed dispersal is itself a non-native, the Japanese white-eye. The berries contain numerous small seeds, one hundred or more in each berry, and once a population is established the seeds may also be spread by being caught up in shoes or the treads of vehicle tires. It is no consolation to know that today this *Clidemia* is also a noxious weed throughout the tropics, in the Old World as well as the New. Even in Central and South America, where it is native, it is known to be invasive in secondary forests. In many places it is known as Koster's curse, a name that appears to have been given it in Fiji. But who this unfortunate person was, or why he or she is attributed with this curse, seems not to be recorded.

Tibouchina urvilleana, as well as being a woody weed, is also an attractive plant with beautiful and striking purple flowers, in contrast with those of *Clidemia hirta*. It is no wonder that as a cultivated plant it has been given the name glorybush. It too comes from South America, in this case from southern Brazil, and it arrived in Hawaii as a garden plant, probably about 1910 near Kurtistown on the Big Island. It had been introduced to Britain for cultivation under glass some time in the middle of the nineteenth century. There, it was and still is greatly admired when in flower, so it is not surprising that sooner or later it was introduced to Hawaii as a garden plant. One thousand plants were even planted in forest

Clidemia hirta (L.) D. Don var. *hirta* and *Tibouchina urvilleana* (DC.) Cogn. var. *urvilleana*
Plants from Kokee, Kauai

reserves on Oahu and Kauai. No one suspected that it would become an ecological menace. But by 1948, F. R. Fosberg reported that it was well established in the upper Nuuanu Valley of Oahu, and he added a solitary plea which, with hindsight, we wish had elicited a response: "It may still be eradicated here at a relatively small cost. Rather than have another insoluble weed problem, it would seem obviously worth while to do this while it is still possible."

In Hawaii, *Tibouchina urvilleana* is now naturalized on the four largest islands, usually at altitudes of between about 1,500 and 4,500 feet, for it prefers moist, relatively cool conditions. It is not so aggressive as the *Clidemia* and, unlike that plant, does not appear to penetrate the forest. Instead, it is generally to be found on the edge of forests and beside roads and tracks, although at lower, drier elevations it seems to prefer some shade.

This plant was long referred to as *Tibouchina semidecandra*, but it has been shown that it should be called *T. urvilleana*, and this is the name to be found in the more recent literature. The epithet *urvilleana* commemorates the French naval officer, hydrographer, and explorer Jules Sébastian César Dumont d'Urville (1790–1842), who commanded several expeditions around the world and collected this plant from the state of Santa Catarina in southern Brazil. The generic name *Tibouchina* was derived from the native name in Guiana for another species in this genus.

SELECTED READINGS

Abbott, Isabella Aiona. *La'au Hawai'i: Traditional Hawaiian Uses of Plants*. Honolulu: Bishop Museum Press, 1992.

Professor Abbott emphasizes the cultural aspects of ethnobotany and introduces the reader to the history and culture of Polynesia, to the plants Polynesians introduced to Hawaii, and to their traditional and contemporary uses.

Bailey, L. H. *How Plants Get Their Names*. New York: Dover, 1963. [Unabridged replication of work first published by Macmillan in 1933.]

This short introduction to the binomial system of nomenclature includes information on Linnaeus, on how plants are identified, and on the rules of nomenclature.

Blunt, Wilfrid, and William T. Stearn. *The Art of Botanical Illustration*. New ed., rev. and enl. Woodbridge, Suffolk: Antique Collectors' Club, 1994.

The undisputed authority on the history of the art of botanical illustration.

Burtt, B. L., and R. M. Smith. "Key Species in the Taxonomic History of the Zingiberaceae." *Notes from the Royal Botanic Garden Edinburgh* 31 (1972):177–227.

Specialist review of the early history and application of names in the ginger family.

Carlquist, Sherwin. *Island Biology*. New York: Columbia University Press, 1974.

Comprehensive, well-illustrated biogeographic investigation of the evolutionary patterns of island plants and animals.

Carlquist, Sherwin. *Hawaii, A Natural History: Geology, Climate, Native Flora and Fauna above the Shoreline*. 2d ed. Lawai, Hawaii: Pacific Tropical Botanical Garden, 1980.

Profusely illustrated with photographs and line drawings, this work provides a comprehensive overview of the unique natural history of the Hawaiian Archipelago.

Carr, Gerald D. "Monograph of the Hawaiian Madiinae (Asteraceae): Argyroxiphium, Dubautia, and Wilkesia." *Allertonia* (Lawai, Kauai) 4 (1985):1–123.

The definitive taxonomic study of the Hawaiian silverswords and their relatives.

Cuddihy, Linda W., and Charles P. Stone. *Alteration of Native Hawaiian Vegetation: Effects of Humans, Their Activities and Introductions*. Honolulu: University of Hawaii, Cooperative National Park Resources Study Unit, 1990.

This extremely readable analysis of the decline of native vegetation in Hawaii since the arrival of humans addresses such topics as native Hawaiian plant communities, the impacts of the aboriginal settlers, early European contact, and modern (post-1850) changes in the vegetation.

Degener, Otto. *Flora Hawaiiensis, or New Illustrated Flora of the Hawaiian Islands*. Honolulu: Published privately, 1932–1969.

Degener came to Hawaii in 1922 and for many years was the Collaborator in Hawaiian Botany for the New York Botanical Garden. His *Flora Hawaiiensis* was published in loose-leaf sets, with information on each plant presented on one page and a drawing on the verso. The sets were intended to remain unbound, so that new sheets could be inserted in their proper place as they were completed. Nonetheless, it is difficult to locate any particular plant without using the indexes.

Degener, Otto. *Plants of Hawaii National Park Illustrative of Plants and Customs of the South Seas*. Honolulu: Published privately, 1983 [originally 1930].

Each plant entry in this work provides information on traditional Hawaiian uses and a botanical description. Includes some interesting early photographs of Hawaiians as well as numerous line drawings of the plants.

Fosberg, F. Raymond. *Immigrant Plants in the Hawaiian Islands*. Occasional Papers, University of Hawaii, Nos. 32, 46. Honolulu: University of Hawaii, 1937, 1948.

These papers provide early documentation of noxious weed introductions, including information on the rates at which various plant pests were spreading.

Fosberg, F. Raymond. "Derivation of the Flora of the Hawaiian Islands." In *Insects of Hawaii*, Vol. 1, ed. E. C. Zimmerman. Honolulu: University of Hawaii Press, 1948.

Still a standard today, this analysis of the Hawaiian flora estimates the number of original ancestors, or introductions, necessary to give rise to the present-day diversity.

Fosberg, F. R., and Derral Herbst. "Rare and Endangered Species of Hawaiian Vascular Plants." *Allertonia* (Lawai, Kauai) 1 (1975):1–72.

This list was originally developed by Fosberg for the first conference on Hawaiian endangered species in 1970. Includes 1,765 Hawaiian plant species, of which 1,172 were classified as extinct, endangered, or rare at the time.

Hillebrand, William. *Flora of the Hawaiian Islands: A Description of Their Phanerogams and Vascular Cryptogams*. Monticello, New York: Lubrecht and Cramer, 1981. [Facsimile of the first edition, London: Williams and Norgate, 1888.]

For over one hundred years, this was the only complete flora of the indigenous and naturalized plants of the Hawaiian Islands.

Kay, E. Alison, ed. *A Natural History of the Hawaiian Islands: Selected Readings II*. Honolulu: University of Hawaii Press, 1994.

This collection of current primary source material examines Hawaii's geological processes, terrestrial environment, marine environment, evolution and endemism, ecosystems, and the effects of humans.

Krauss, Beatrice H. *Plants in Hawaiian Culture*. Honolulu: University of Hawaii Press, 1993.

A well-known local ethnobotanist, Dr. Krauss provides a readable overview of Hawaiian ethnobotany. The focus of this book is on the relationship between plants and pre-contact Hawaiian culture.

Lowe, David W., John R. Matthews, and Charles J. Moseley, eds. *The Official World Wildlife Fund Guide to Endangered Species of North America*. 4 vols. Washington, D.C.: Beacham, 1990–94.

These volumes provide a comprehensive guide to continental U.S., Alaskan, Hawaiian, and Puerto Rican species protected by the Endangered Species Act as of July 1994. The entry for each species includes a photograph and description, information on habitat, current conservation status, and bibliography.

Mueller-Dombois, D., K. W. Bridges, and H. L. Carson. *Island Ecosystems: Biological Organization in Selected Hawaiian Communities*. US/IBP Synthesis Series 15. Woods Hole, Massachusetts: Hutchinson Ross, 1981.

The papers in this volume synthesize the work of researchers studying the biological organization in selected ecosystems of the Hawaiian Islands.

Neal, Marie C. *In Gardens of Hawaii*. Rev. ed. Bernice P. Bishop Museum Special Publication 50. Honolulu: Bishop Museum Press, 1965.

Written for the horticulturalist and gardener rather than the professional botanist, Neal's book covers over 3,000 species of ornamentals, fruits, vegetables, weeds, and some native species found in Hawaii's gardens. *In Gardens of Hawaii* was first published in 1948 and, though much of the taxonomy has been superseded, it is still the only reasonably comprehensive work on this subject. A thoroughly revised and rewritten version is now in preparation at the Bishop Museum, Honolulu.

Purseglove, J. W. "The Spread of Tropical Crops." In *The Genetics of Colonizing Species: Proceedings of the First International Union of Biological Sciences Symposia on General Biology*, ed. H. G. Baker and G. Ledyard Stebbins. New York: Academic Press, 1965.

In this review of the early history of intercontinental crop distribution, Professor Purseglove analyses the thesis that humans brought the sweet potato from South America to Polynesia.

Report on the Rare Plants of Hawaii. St. Louis: Missouri Botanical Garden, Center for Plant Conservation, 1992.

This list was developed by the Center for Plant Conservation and the Rare Plant Task Force in Hawaii to determine priorities for germplasm storage and *ex situ* cultivation. Over four hundred species are identified as being of conservation concern, including 165 species with one hundred or fewer individuals and 105 species that are considered extinct or likely to be extinct.

Rock, Joseph F. *A Monographic Study of the Hawaiian Species of the Tribe Lobelioideae Family Campanulaceae*. Memoirs of the Bernice P. Bishop Museum 7, no. 4 (1919).

Classic study, with extensive photographs, of this most unique family of Hawaiian plants.

Rock, Joseph F. *The Indigenous Trees of the Hawaiian Islands*. Reprinted with introduction and addenda by the Pacific Tropical Botanical Garden, Lawai, Hawaii. Rutland, Vermont: Charles E. Tuttle, 1974. [Originally published privately, 1913.]

This volume on the native trees of Hawaii is particularly notable for Rock's exceptional black-and-white photographs, which depict some of the finest specimens of Hawaii's native trees and of the natural plant communities existing at the beginning of this century.

Stone, Charles P., and Linda W. Pratt. *Hawai'i's Plants and Animals: Biological Sketches of Hawaii Volcanoes National Park*. Honolulu: Hawaii Natural History Association, 1994.

Hawaii Volcanoes National Park, a World Heritage Site and International Biosphere Reserve, encompasses a continuous range of ecological zones from seaside up to alpine/aeolian. This popular field guide to the Park includes discussions and line drawings of both native and introduced fauna and flora.

Stone, Charles P., and J. Michael Scott, eds. *Hawai'i's Terrestrial Ecosystems: Preservation and Management*. Proceedings of a symposium held June 5–6, 1984, at Hawaii Volcanoes National Park. Honolulu: University of Hawaii, Cooperative National Park Resources Studies Unit, 1985.

This proceedings publication includes practical as well as scientific information on a range of topics related to preservation of habitat and protection of endangered species in Hawaii. Chapters are included on habitat status assessment, status of native flowering plants, impacts of alien plants and animals, and ecosystem monitoring, management, and restoration.

Stone, Charles P., Clifford W. Smith, and J. Timothy Tunison, eds. *Alien Plant Invasions in Native Ecosystems in Hawai'i: Management and Research*. Honolulu: University of Hawaii, Cooperative National Park Resources Studies Unit, 1992.

Papers originally presented in a 1986 symposium were updated for this collection. The paper by Lyndon Wester, "Origin and Distribution of Adventive Flowering Plants in Hawaii," includes a 37-page listing of "Alien Adventive Seed Plants in the Hawaiian Islands."

Wagner, Warren L., Derral R. Herbst, and S. H. Sohmer. *Manual of the Flowering Plants of Hawai'i*. 2 vols. Bishop Museum Special Publication 83. Honolulu: University of Hawaii Press and Bishop Museum Press, 1990.

These two volumes on the native and naturalized flowering plants of the Hawaiian Islands are the first completely updated comprehensive manual on the subject in over a century. As well as keys and descriptions of families, genera, and species based on taxonomic literature through September 1987, this work includes detailed reviews of Island geology, climate, and plant communities There are numerous line drawings, at least one for each genus, and ethnobotanical information is included for many of the species.

PLANT INDEX

'a'ali'i 24
Acacia heterophylla 8
Acacia koa 8
Acacia simplex 8
'āhinahina 6, 22, 24, 40
aka'aka'awa 10
'ākala 20
'ākalakala 20
Aleurites moluccana 60
Alpinia nutans 82
Alpinia speciosa 82
Alpinia zerumbet 82
ālula 50
'ānaunau 2, 44
angel's trumpet 72
Argyroxiphium grayanum 6, 22, 40
Argyroxiphium sandwicense 6, 22, 40
Argyroxiphium virescens 6, 22
Aristolochia clematitis 84
Aristolochia elegans 84
Aristolochia littoralis 84
Artocarpus altilis 64, 66
Artocarpus heterophyllus 86
Averrhoa carambola 80
'awapuhi luheluhe 82

banana poka 92, 94
bat flower 70
beach naupaka 32
bird of paradise 74
birthwort 84
breadfruit 64, 66, 86
Brighamia citrina 50
Brighamia insignis 50
Brighamia rockii 50
Brugmansia aurea 72
Brugmansia versicolor 72
Brugmansia × candida 72

Caesalpinia bonduc 36
Caesalpinia decapetala 36
Caesalpinia kavaiensis 36
Caesalpinia major 36
Caesalpinia scortechinii 36
calico flower 84
candlenut tree 60
carambola 80
Catimbium speciosum 82
Cereus undatus 76
chempedak 86
Clidemia hirta 96, 98
Colocasia esculenta 58, 64
Cordyline fruticosa 18, 58, 64
crane flower 74
Crinum amabile 78
Crinum augustum 78
Crinum pedunculatum 78
Cyanea coriacea 46

Datura arborea 72
Datura candida 72
devil flower 70
Dodonaea eriocarpa 24
Dodonaea viscosa 24
Dracaena aurea 26

Erythrina sandwicensis 2

Florida blackberry 20
floripondio 72
fountain grass 36

garden cress 44
giant granadilla 94
glorybush 96
golden dracaena 26
greensword 6, 22, 40

hāhā 46, 50
hairy merremia 88
hala pepe 26
hau 56
Hibiscus brackenridgei 48
Hibiscus kahilii 48
Hibiscus kokio 38, 48
Hibiscus rosa-sinensis 48
Hibiscus tiliaceus 48, 56
Hillebrandia sandwicensis 10
hona 18
hōpue 18
huahekili 32
Hylocereus undatus 76
'ihi 50
'iliahi 14
iliau 6, 22, 24
Indian lotus 90
Ipomoea batatas 62
Ipomoea purga 88
Ipomoea tiliacea 62
Ipomoea trifida 62

jack fruit 86

ka'e'e 68
kala 20
kalo 58
ki 18
koa 8, 64
Kokia cookei 38
Kokia drynarioides 38
Kokia kauaiensis 12, 38
Kokia lanceolata 38
koki'o 12, 38
koki'o 'ula'ula 48
Koster's curse 96
kukui 24, 60, 64
kū makani 24
kumar 62
kumara 62

lā'au 'ala 14
lehua 16
lehua mamo 16
Lepidium arbuscula 44
Lepidium bidentatum 44
Lepidium sativum 44
Lepidium serra 2, 44
liliko'i 94
Lobelia erinus 12
Lobelia gaudichaudii 12
Lobelia gloria-montis 12
Lobelia kauaensis 12
lotus bean 90
loulu 52

mākū 48
Mauritius raspberry 20
Merremia aegyptia 88

Merremia tuberosa 88
Metrosideros collina 16
Metrosideros polymorpha 16
Mezoneuron kavaiensis 36
monkey pod tree 70
morning glory 62
mountain naupaka 32
Mucuna bennettii 68
Mucuna gigantea 68
Mucuna novoguineënsis 68
Mucuna urens 68
Munroidendron racemosum 42
Myoporum sandwicense 14

nānāhonua 72
native red cotton 38
naupaka kahakai 32
naupaka kuahiwi 32
Nelumbo lutea 90
Nelumbo nucifera 58, 90
New Guinea creeper 68
night-blooming cereus 76

ōhāhā 50
'ohelo 30
'ohe naupaka 12
'ōhia 16
'ōhia lehua 2, 16
olomea 56
olonā 18
'ōlulu 50
ōpuhe 18
ōulu 50

pāmakani 28
pānini-o-Ka-puna-hou 76
Passiflora edulis 94
Passiflora ligularis 94
Passiflora maliformis 94
Passiflora mixta 92
Passiflora mollissima 92
Passiflora quadrangularis 94
passion flower 94
Pennisetum setaceum 36
Perrottetia sandwicensis 56
pia 70
pilikai 88
Pleomele aurea 26
Polynesian arrowroot 70
Portulaca molokiniensis 50
Pritchardia affinis 52
Pritchardia aylmer-robinsonii 52
Pritchardia eriophora 52
Pritchardia hardyi 52
Pritchardia kaalae 52
Pritchardia minor 52
Pritchardia munroi 52, 54
Pritchardia napaliensis 52
Pritchardia remota 52
Pritchardia schattaueri 52
Pritchardia viscosa 52
pua 'ala 50
pua maka nui 10
pū aupaka 50
pu'e 12

Queen Emma lily 78

red jade vine 68
Rubus argutus 20
Rubus hawaiensis 20
Rubus macraei 20
Rubus penetrans 20
Rubus rosifolius 20

sacred lotus 90
Samanea saman 70
sandalwood 14
Santalum freycinetianum 14
Scaevola gaudichaudiana 32
Scaevola glabra 12
Scaevola mollis 32
Scaevola procera 32
Scaevola sericea 32
Scaevola taccada 32
sea bean 68
shell flower 82
shell ginger 82
silversword 6, 22, 40
starfruit 80
Strelitzia reginae 74
sweet calabash 94
sweet granadilla 94
sweet potato 62

Tacca chantrierorum 70
Tacca leontopetaloides 70
taro 36, 58, 62, 64
thimbleberry 20
ti 18, 58, 64
Tibouchina semidecandra 98
Tibouchina urvilleana 96, 98
Touchardia latifolia 18

'uala 62
uhiuhi 36
'ulu 64
Urera glabra 18
Urera sandwicensis 18
'uwala 62

Vaccinium cereum 30
Vaccinium peleanum 30
Vaccinium reticulatum 30
Viola chamissoniana 28
Viola tracheliifolia 28

wiliwili 2
Wilkesia gymnoxiphium 6
Wilkesia hobdyi 6
wood rose 88

Zerumbet speciosum 82
Zingiber zerumbet 82
Zizyphus lotus 90